RAIN WATER HARVESTING

RAIN WATER HARVESTING

by

Navjot Bera | Chirag Badal | Sarabjit Borra

Wordpen Academics

Publishing Science That Matter

ISBN: 978-93-6084-008-2

Published by : **WORDPEN ACADEMICS**
4736/23, Ansari Road, Darya Ganj,
New Delhi – 110 002
Phone: +91-011-35004336
E-mail: info@wordpenacademics.com

Printed at **:** **Replika Press Pvt. Ltd.**

Exclusive Distributor: ***Bio-Green Books, New Delhi***

PREFACE

"Rainwater Harvesting" delves into the crucial practice of collecting and storing rainwater for various purposes, offering readers a comprehensive understanding of its importance, methods, benefits, and applications. This book serves as an essential resource for individuals, communities, policymakers, and environmentalists seeking sustainable solutions to water scarcity, climate change, and resource management challenges.

Water scarcity is a pressing global issue exacerbated by factors such as population growth, urbanization, climate change, and inefficient water management practices. Rainwater harvesting presents a sustainable and cost-effective solution to alleviate water stress by harnessing the abundant and renewable resource of rainwater that falls from the sky.

The primary objective of this book is to provide readers with a thorough exploration of rainwater harvesting techniques, ranging from simple household systems to large-scale infrastructure projects. Each chapter offers insights, practical guidance, case studies, and best practices to help readers implement rainwater harvesting systems effectively and responsibly.

Contributions from experts in the field cover various aspects of rainwater harvesting, including rooftop rainwater collection, storage tanks and cisterns, filtration and purification methods, water quality monitoring, irrigation systems, and policy frameworks. Special emphasis is placed on the environmental, social, and economic benefits of rainwater harvesting, such as water conservation, groundwater recharge, drought resilience, and community empowerment.

As we confront the challenges of water scarcity and climate change, it is imperative to recognize the importance of decentralized and nature-based solutions like rainwater harvesting. By harnessing the power of rainwater, individuals and communities can reduce reliance on centralized water sources, alleviate pressure on water infrastructure, and promote environmental sustainability.

We extend our sincere gratitude to the contributors whose expertise and dedication have enriched this book. It is our hope that "Rainwater Harvesting" will serve as a valuable resource and catalyst for action, inspiring readers to embrace rainwater harvesting as a vital component of sustainable water management strategies for a more resilient and water-secure future.

Navjot Bera | Chirag Badal | Sarabjit Borra

CONTENTS

1

INTRODUCTION

Rainwater harvesting provides an independent water supply during regional water restrictions and in developed countries is often used to supplement the main supply. It provides water when there is a drought, can help mitigate flooding of low-lying areas, and reduces demand on wells which may enable groundwater levels to be sustained. It also helps in the availability of potable water as rainwater is substantially free of salinity and other salts. Application of rainwater harvesting in urban water system provides a substantial benefit for both water supply and wastewater subsystems by reducing the need for clean water in water distribution system, less generated stormwater in sewer system, as well as a reduction in stormwater runoff polluting freshwater bodies.

There has been a large body of work focused on the development of Life Cycle Assessment and Life Cycle Costing methodologies to assess the level of environmental impacts and money that can be saved by implementing rainwater harvesting systems.

More development and knowledge is required to understand the benefits rainwater harvesting can provide to agriculture. Many countries especially those with an arid environment use rainwater harvesting as a cheap and reliable source of clean water. To enhance irrigation in arid environments, ridges of soil are constructed in order to trap and prevent rainwater from running down hills and slopes. Even in periods of low rainfall, enough water is collected in order for crops to grow. Water can be collected from roofs, dams, and ponds can be constructed in order to hold large quantities of rainwater so that even on days where there is little to no rainfall, there is enough available to irrigate crops.

Quality

The concentration of contaminants is reduced significantly by diverting the initial flow of run-off water to waste. Improved water quality can also be obtained by using a floating draw-off mechanism (rather than from the base of the tank) and by using a series of tanks, with draw from the last in series. Pre-filtration is a common practice used in the industry to ensure that the water entering the tank is free of large sediment. Pre-filtration is important to keep the system healthy.

Conceptually, a water supply system should match the quality of water with the end use. However, in most of the developed world high quality potable water is used for all end uses. This approach wastes money and energy and imposes unnecessary impacts to the environment. Supplying rainwater that has gone through preliminary filtration measures for non-potable water uses, such as toilet flushing, irrigation, and laundry, may be a significant part of a sustainable water management strategy.

System Setup

Rainwater harvesting systems can range in complexity, from systems that can be installed with minimal skills, to automated systems that require advanced setup and installation. The basic Rainwater harvesting system is more of a plumbing job than a technical job as all the outlets from the building terrace are connected through a pipe to an underground tank that stores water.

Systems are ideally sized to meet the water demand throughout the dry season since it must be big enough to support daily water consumption. Specifically, the rainfall capturing area such as a building roof must be large enough to maintain adequate flow. The water storage tank size should be large enough to contain the captured water. For low-tech systems, there are many low-tech methods used to capture rainwater: rooftop systems, surface water capture, and pumping the rainwater that has already soaked into the ground or captured in reservoirs and storing it into tanks (cisterns).

Before a rainwater harvesting system is built, it's helpful to use digital tools. For instance, if you want to detect if a region has a high rainwater harvesting potential, rainwater harvesting GIS maps can be made using an online interactive tool. Or if you need to estimate how much water is needed to fulfill a community's water needs, the Rain is Gain tool helps with this. Tools like these can save time and money before a commitment to build a system is undertaken, in addition to making the project sustainable and last a long time.

Life Cycle Assessment: Design for Environment

Contemporary system designs require an analysis of not only the economic and technical performance of a system, but also the environmental performance. Life

Cycle Assessment is a methodology used to evaluate the environmental impacts of a precut or systems, from cradle-to-grave of its' lifetime. Devkota et al., developed such a methodology for rainwater harvesting, and found that the building design (e.g., dimensions) and function (e.g., educational, residential, etc.) play critical roles in the environmental performance of the system. The Economic and Environmental Analysis of Sanitations Technologies, EEAST model evaluates the greenhouse gas emissions and cost of such systems over the lifetime of a variety of building types.

To address the functional parameters of rainwater harvesting systems, a new metric was developed - the demand to supply ratio (D/S) - identifying the ideal building design (supply) and function (demand) in regard to the environmental performance of rainwater harvesting for toilet flushing. With the idea that supply of rainwater not only saves the potable water, but also saves the stormwater entering the combined sewer network (thereby requiring treatment), the savings in environmental emissions were higher if the buildings are connected to a combined sewer network compared to separate one.

Rain water harvesting by freshwater flooded forests

Rain water harvesting is possible by growing fresh water flooded forests without losing the income from the used /submerged land. The main purpose of the rain water harvesting is to utilize the locally available rain water to meet water requirements throughout the year without the need of huge capital expenditure. This would facilitate availability of uncontaminated water for domestic, industrial and irrigation needs.

New approaches

Instead of using the roof for catchment, the RainSaucer, which looks like an upside down umbrella, collects rain straight from the sky. This decreases the potential for contamination and makes potable water for developing countries a potential application. Other applications of this free standing rainwater collection approach are sustainable gardening and small plot farming.

A Dutch invention called the Groasis Waterboxx is also useful for growing trees with harvested and stored dew and rainwater.

Traditionally, storm water management using detention basins served a single purpose. However, Optimized Real-Time Control (OptiRTC) lets this infrastructure double as a source of rainwater harvesting without compromising the existing detention capacity. This has been used in the EPA headquarters to evacuate stored water prior to storm events, thus reducing wet weather flow while ensuring water availability for later reuse. This has the benefit of increasing water quality released and decreasing the volume of water released during combined sewer overflow events.

Generally, check dams are constructed across the streams to enhance the percolation of surface water in to the sub soil strata. The water percolation in the water impounded area of the check dams, can be enhanced artificially many folds by loosening the sub soil strata / overburden by using ANFO explosives as used in open cast mining. Thus local aquifers can be recharged quickly by using the available surface water fully for using in the dry season.

HISTORY

Earlier

Around the third century BC, the farming communities in Balochistan (now located in Pakistan, Afghanistan and Iran), and Kutch, India, used rainwater harvesting for agriculture and many uses also. In ancient Tamil Nadu, rainwater harvesting was done by Chola kings. Rainwater from the Brihadeeswarar temple (located in Balaganpathy Nagar, Thanjavur, India) was collected in Shivaganga tank. During the later Chola period, the Vîrânam tank was built (1011 to 1037 CE) in Cuddalore district of Tamil Nadu to store water for drinking and irrigation purposes. Vîrânam is a 16 km (9.9 mi) long tank with a storage capacity of 1,465,000,000 cubic feet (41,500,000 m).

Rainwater harvesting was done in the Indian states of Madhya Pradesh, Maharashtra, and Chhattisgarh in the olden days. Ratanpur, in the state of Chhattisgarh, had around 150 ponds. Most of the tanks or ponds were utilized in agriculture works.

Present day

- Currently in China and Brazil rooftop rainwater harvesting is being practiced for providing drinking water, domestic water, water for livestock, water for small irrigation and a way to replenish groundwater levels. Gansu province in China and semi-arid north east Brazil have the largest rooftop rainwater harvesting projects ongoing.
- In Bermuda, the law requires all new construction to include rainwater harvesting adequate for the residents.
- The U.S. Virgin Islands have a similar law.
- In Senegal and Guinea-Bissau, the houses of the Diola-people are frequently equipped with homebrew rainwater harvesters made from local, organic materials.
- In the Irrawaddy Delta of Myanmar, the groundwater is saline and communities rely on mud-lined rainwater ponds to meet their drinking

water needs throughout the dry season. Some of these ponds are centuries old and are treated with great reverence and respect.

- In the United States: until 2009 in Colorado, water rights laws almost completely restricted rainwater harvesting; a property owner who captured rainwater was deemed to be stealing it from those who have rights to take water from the watershed. Now, residential well owners that meet certain criteria may obtain a permit to install a rooftop precipitation collection system (SB 09-080). Up to 10 large scale pilot studies may also be permitted (HB 09-1129). The main factor in persuading the Colorado Legislature to change the law was a 2007 study that found that in an average year, 97% of the precipitation that fell in Douglas County, in the southern suburbs of Denver, never reached a stream—it was used by plants or evaporated on the ground. In Colorado you cannot even drill a water well unless you have at least 35 acres (14 ha). In New Mexico, rainwater catchment is mandatory for new dwellings in Santa Fe. Texas offers a sales tax exemption on the purchase of rainwater harvesting equipment. Both Texas and Ohio allow the practice even for potable purposes. Oklahoma passed the Water for 2060 Act in 2012, to promote pilot projects for rainwater and graywater use among other water saving techniques.
- In Beijing, some housing societies are now adding rain water in their main water sources after proper treatment.
- In Ireland, Professor MichealMcginley established a project to design a rain water harvesting prototype in the Biosystems design Challenge Module at University College Dublin.

Canada

A number of Canadians have started implementing rainwater harvesting systems for use in stormwater reduction, irrigation, laundry, and lavatory plumbing. Substantial reform to Canadian law since the mid 2000s has increased the use of this technology in agricultural, industrial, and residential use; but ambiguity remains amongst legislation in many provinces. Bylaws and local municipal codes often regulate rainwater harvesting.

India

- Karnataka: In Bangalore it is mandatory for adoption of rain water harvesting for every owner or the occupier of a building having the sital area measuring 60 ft (18.3 m) X 40 ft (12.2 m) and above and for newly constructed building measuring 30 ft (9.1 m) X 40 ft (12.2 m) and above dimension. In this regard BWSSB has initiated and constructed "Rain Water Harvesting Theme Park" in the name of Sir. M. Visvesvaraya in 1.2 acres (4,900 m) land situated at Jayanagar, Bangalore. In this park 26 different type of Rain Water Harvesting models are demonstrated along with the water conservation tips. The Auditorium on the first floor is set up with Green Air conditioning system and the same will be utilized to arrange the meeting and showing the video clip about the rain water harvesting to students as well as general public.

- Tamil Nadu: In the state of Tamil Nadu, rainwater harvesting was made compulsory for every building to avoid groundwater depletion. It gave excellent results within five years, and every state took it as role model. Since its implementation, Chennai saw a 50 percent rise in water level in five years and the water quality significantly improved.
- Rajasthan: In Rajasthan, rainwater harvesting has traditionally been practiced by the people of the Thar Desert. There are many ancient water harvesting systems in Rajasthan, which have now been revived. Water harvesting systems are widely used in other areas of Rajasthan as well, for example the chauka system from the Jaipur district.
- Maharashtra: At present, in Pune, rainwater harvesting is compulsory for any new housing society to be registered.
- In Mumbai city in Maharashtra rain water harvesting is being considered as a good solution to solve water crisis.

The Mumbai city council is planning to make rainwater harvesting mandatory for large societies.

- An attempt has been made at the Department of Chemical Engineering, IISc, Bangalore to harvest rainwater using upper surface of a solar still, which was used for water distillation

Israel

- The Southwest Center for the Study of Hospital and Healthcare Systems in cooperation with Rotary International is sponsoring a rainwater harvesting model program across the country. The first rainwater catchment system was installed at an elementary school in Lod, Israel. The project is looking to expand to Haifa in its third phase. The Southwest Center has also partnered with the Water Resources Action Project

(WRAP) of Washington D.C. WRAP currently has rainwater harvesting projects in the West Bank.

- Rainwater harvesting systems are being installed in local schools for the purpose of educating schoolchildren about water conservation principles and bridging divides between people of different religious and ethnic backgrounds all while addressing the water scarcity issue that the Middle East faces.

Sri Lanka

- Rainwater harvesting has been a popular method of obtaining water for agriculture and for drinking purposes in rural homes.
- The legislation to promote rainwater harvesting was enacted through the Urban Development Authority (Amendment) Act, No. 36 of 2007.
- Lanka rainwater harvesting forum is leading the Sri Lanka's initiative.

South Africa

- The South African Water Research Commission has supported research into rainwater harvesting. Reports on this research are available on their 'Knowledge Hub'.
- Studies in arid, semi-arid and humid regions have confirmed that techniques such as mulching, pitting, ridging and modified run-on plots are effective for small-scale crop production.

United Kingdom

In the United Kingdom, water butts are often found in domestic gardens to collect rainwater, which is then used to water the garden. However, the British government's Code For Sustainable Homes encourages fitting large underground tanks to new-build homes to collect rainwater for flushing toilets, washing clothes, watering the garden, and washing cars. This reduces by 50 percent the amount of mains water used by the home.

United States

- Colorado: Until 2009, water rights laws almost completely restricted rainwater harvesting; a property owner who captured rainwater was deemed to be stealing it from those who have rights to take water from the watershed. Now, residential well owners that meet certain criteria may obtain a permit to install a rooftop precipitation collection system (SB 09-080). Up to 10 large scale pilot studies may also be permitted (HB

09-1129). The main factor in persuading the Colorado Legislature to change the law was a 2007 study that found that in an average year, 97% of the precipitation that fell in Douglas County, in the southern suburbs of Denver, never reached a stream—it was used by plants or evaporated on the ground. In Colorado you cannot even drill a water well unless you have at least 35 acres (14 ha)

- New Mexico: In New Mexico, rainwater catchment is mandatory for new dwellings in Santa Fe.
- Texas offers a sales tax exemption on the purchase of rainwater harvesting equipment. Both Texas and Ohio allow the practice even for potable purposes.
- Oklahoma passed the Water for 2060 Act in 2012, to promote pilot projects for rainwater and graywater use among other water saving techniques.

Non-traditional

- In 1992, American artist Michael Jones McKean created an artwork in Omaha, Nebraska at the Bemis Center for Contemporary Art that created a fully sustainable rainbow in the Omaha skyline. The project collected thousands of gallons of rainwater storing the water in six daisy-chained 12,000 gallons tanks. The massive logistical undertaking, during its 5-month span, was one of the largest urban rainwater harvesting sites in the American Mid-West.

PEAK WATER

Peak water is a concept that underlines the growing constraints on the availability, quality, and use of freshwater resources.

Peak water is defined in a 2010 peer-reviewed article in the Proceedings of the National Academy of Sciences by Peter Gleick and MeenaPalaniappan. They distinguish between peak renewable, peak non-renewable, and peak ecological water in order to demonstrate the fact that although there is a vast amount of water on the planet, sustainably managed water is becoming scarce.

Lester R. Brown, president of the Earth Policy Institute, wrote in 2013 that although there was extensive literature on peak oil, it was peak water that is "the real threat to our future". An assessment was published in August 2011 in the Stockholm International Water Institute's journal. Much of the world's water in underground aquifers and in lakes can be depleted and thus resembles a finite resource. The phrase peak water sparks debates similar to those about peak oil. In 2010, New York Times chose "peak water" as one of its 33 "Words of the Year".

There are concerns about impending peak water in several areas around the world:

- Peak renewable water, where entire renewable flows are being consumed for human use
- Peak non-renewable water, where groundwater aquifers are being overpumped (or contaminated) faster than nature recharges them (this example is most like the peak oil debate)
- Peak ecological water, where ecological and environmental constraints are overwhelming the economic benefits provided by water use

If present trends continue, 1.8 billion people will be living with absolute water scarcity by 2025, and two-thirds of the world could be subject to water stress. Ultimately, peak water is not about running out of freshwater, but about reaching physical, economic, and environmental limits on meeting human demands for water and the subsequent decline of water availability and use.

COMPARISON WITH PEAK OIL

The Hubbert curve has become popular in the scientific community for predicting the depletion of various natural resources. M. King Hubbert created this measurement device in 1956 for a variety of finite resources such as coal, oil, natural gas and uranium. Hubbert's curve was not applied to resources such as water originally, since water is a renewable resource. Some forms of water, however, such as fossil water, exhibit similar characteristics to oil, and overpumping (faster than the rate of natural recharge of groundwater) often results in a Hubbert-type peak. A modified Hubbert curve applies to any resource that can be harvested faster than it can be replaced. Like peak oil, peak water is inevitable given the rate of extraction of certain water systems. A current argument is that growing populations and demands for water will inevitably lead to non-renewable use of water resources.

Water supply

Fresh water is a renewable resource, yet the world's supply of clean, fresh water is under increasing demand for human activities. The world has an estimated 1.34 billion cubic kilometers of water, but 96.5% of it is salty. Almost 70% of fresh water can be found in the ice caps of Antarctica and Greenland. Less than 1% of this water on Earth is accessible to humans, the rest is contained in soil moisture or deep underground. Accessible freshwater is located in lakes, rivers, reservoirs and shallow underground sources. Rainwater and snowfall do very little to replenish these stocks of freshwater.The amount of available freshwater supply in some regions is decreasing because of (i) climate change, which has caused receding glaciers, reduced stream and river flow, and shrinking lakes; (ii) contamination of water by human and industrial wastes; and (iii) overuse of non-renewable groundwater aquifers. Many aquifers have been over-pumped and are not recharging quickly. Although the total freshwater supply is not used up, much has become polluted, salted, unsuitable or otherwise unavailable for drinking, industry and agriculture.

Water demand

Water demand already exceeds supply in many parts of the world, and as the world population continues to rise, many more areas are expected to experience this imbalance in the near future. Agriculture represents 70% of freshwater use worldwide.

Agriculture, industrialization and urbanization all serve to increase water consumption.

Freshwater withdrawal by country

The highest total annual consumption of water comes from India, China and the United States, countries with large populations, extensive agricultural irrigation, and demand for food.

India

India has 20 percent of the Earth's population, but only four percent of its water. Water tables are dropping rapidly in some of India's main agricultural areas. The Indus and Ganges rivers are tapped so heavily that, except in rare wet years, they no longer reach the sea.

India has the largest water withdrawal out of all the countries in the world. Eighty-six percent of that water supports agriculture. That heavy use is dictated in large part by what people eat. People in India consume a lot of rice. Rice farmers in India typically get less than half the yield per unit area while using ten times more water than their Chinese counterparts. Economic development can make things worse because as people's living standards rise, they tend to eat more meat, which requires lots of water to produce. Growing a tonne of grain requires 1,000 tonnes of water; producing a tonne of beef requires 15,000 tonnes. To make a single hamburger requires around 4,940 liters (1,300 gallons) of water A glass of orange juice needs 850 liters (225 gallons) of freshwater to produce.

China

China, the world's most populous country, has the second largest water withdrawal; 68% supports agriculture while its growing industrial base consumes 26%. China is facing a water crisis where water resources are overallocated, used inefficiently, and severely polluted by human and industrial wastes. One-third of China's population lacks access to safe drinking water. Rivers and lakes are dead and dying, groundwater aquifers are over-pumped, uncounted species of aquatic life have been driven to extinction, and direct adverse impacts on both human and ecosystem health are widespread and growing.

In western China's Qinghai province, through which the Yellow River's main stream flows, more than 2,000 lakes have disappeared over the last 20 years. There were once 4,077 lakes. Global climate change is responsible for the reduction in flow of the (Huang He) Yellow River over the past several decades. The source of the Yellow River is the Qinghai-Xizang Tibetan Plateau where the glaciers are receding sharply.

In Hebei Province, which surrounds Beijing, the situation is much worse. Hebei is one of China's major wheat and corn growing provinces. The water tables have been falling fast throughout Hebei. The region has lost 969 of its 1,052 lakes. About 500,000 people are affected by a shortage of drinking water due to continuing droughts. Hydro-power generation is also impacted. Beijing and Tianjin depend on Hebei Province to supply their water from the Yangtze River. Beijing gets its water via the newly constructed South-North Water Transfer Project. The river originates in a glacier on the eastern part of the Tibetan Plateau.

United States

The United States has about 5% of the world's population, yet it uses almost as much water as India (~1/5 of world) or China (1/5 of world) because substantial amounts of water are used to grow food exported to the rest of the world. The U.S. agricultural sector consumes more water than the industrial sector, though substantial quantities of water are withdrawn (but not consumed) for power plant cooling systems. 40 out of 50 state water managers expect some degree of water stress in their state in the next 10 years.

The Ogallala Aquifer in the southern high plains (Texas and New Mexico) is being mined at a rate that far exceeds replenishment—a classic example of peak non-renewable water. Portions of the aquifer will not naturally recharge due to layers of clay between the surface and the water-bearing formation, and because rainfall rates simply do not match rates of extraction for irrigation. The term fossil water is sometimes used to describe water in aquifers that was stored over centuries to millennia. Use of this water is not sustainable when the recharge rate is slower than the rate of groundwater extraction.

In California, massive amounts of groundwater are also being withdrawn from Central Valley groundwater aquifers—unreported, unmonitored, and unregulated. California's Central Valley is home to one-sixth of all irrigated land in the United States, and the state leads the nation in agricultural production and exports. The inability to sustain groundwater withdrawals over time may lead to adverse impacts on the region's agricultural productivity.

The Central Arizona Project (CAP) is a 336-mile (541 km) long canal that diverts 489 billion US gallons (1.85×10 m) a year from the Colorado River to irrigate more than 300,000 acres (1,200 km) of farmland. The CAP project also provides drinking water for Phoenix and Tucson. It has been estimated that Lake Mead, which dams the Colorado, has a 50-50 chance of running dry by 2021.

The Ipswich River near Boston now runs dry in some years due to heavy pumping of groundwater for irrigation. Maryland, Virginia and the District of Columbia have been fighting over the Potomac River. In drought years like 1999 or 2003, and on hot summer days the region consumes up to 85 percent of the river's flow.

PER CAPITA WITHDRAWAL OF WATER

Turkmenistan

Turkmenistan gets most of its water from the Amu Darya River. The Qaraqum Canal is a canal system that takes water from the Amu Darya River and distributes the water out over the desert for irrigation of its orchard crops and cotton. Turkmenistan uses the most water per capita in the world because only 55% of the water delivered to the fields actually reaches the crops.

Kazakhstan and Uzbekistan

The two rivers feeding the Aral Sea were dammed up and the water was diverted to irrigate the desert so that cotton could be produced. As a result, the Aral Sea's water has become much saltier and the sea's water level has decreased by over 60%. Drinking water is now contaminated with pesticides and other agricultural chemicals and contains bacteria and viruses. The climate has become more extreme in the area surrounding it.

Saudi Arabia

According to Walid A. Abderrahman (2001), "Water Demand Management in Saudi Arabia", Saudi Arabia reached peak water in the early 1990s, at more than

30 billion cubic meters per year, and declined afterward. The peak had arrived at about midpoint, as expected for a Hubbert curve. Today, the water production is about half the peak rate. Saudi Arabian food production has been based on "fossil water"—water from ancient aquifers that is being recharged very slowly, if at all. Like oil, fossil water is non-renewable, and it is bound to run out someday. Saudi Arabia has abandoned its self-sufficient food production and is now importing virtually all of its food. Saudi Arabia has built desalination plants to provide about half the country's freshwater. The remainder comes from groundwater (40%), surface water (9%) and reclaimed wastewater (1%).

Libya

Libya is working on a network of water pipelines to import water, called the Great Manmade River. It carries water from wells tapping fossil water in the Sahara desert to the cities of Tripoli, Benghazi, Sirte and others. Their water also comes from desalination plants.

Yemen

Peak water has occurred in Yemen. Sustainability is no longer attainable in Yemen, according to the government's five-year water plan for 2005–2009. The aquifer that supplies Sana'a, the capital of Yemen, will be depleted by 2009. In its search for water in the basin, the Yemeni government has drilled test wells that are 2 kilometers (1.2 mi) deep, depths normally associated with the oil industry, but it has failed to find water. Yemen must soon choose between relocating the city and building a pipeline to coastal desalination plants. The pipeline option is complicated by Sana'a's altitude of 2,250 m (7,380 ft). As of 2010, the threat of running out of water was considered greater than that of Al-Qaeda or instability. There was speculation that Yemenis would have to abandon mountain cities, including Sana'a, and move to the coast. The cultivation of khat and poor water regulation by the government were partly blamed.

United Arab Emirates

United Arab Emirates has a rapidly growing economy and very little water to support it. UAE requires more water than is naturally available. They have reached peak water. To solve this, UAE has a desalination plant near Ruwais and ships its water via pipeline to Abu Dhabi.

CONSEQUENCES

Famine

Water shortage may cause famine in Pakistan. Pakistan has approximately 35 million acres (140,000 km) of arable land irrigated by canals and tube wells, mostly using water from the Indus River. Dams were constructed at Chashma, Mangla, and Tarbela to feed the irrigation system. Since the completion of the Tarbela Dam in 1976 no new capacity has been added despite astronomical growth in population. The gross capacity of the three dams has decreased because of sedimentation, a continual process. Per-capita surface-water availability for irrigation was 5,260 cubic meters per year in 1951. This has been reduced to a mere 1,100 cubic meters per year in 2006.

Health problems

The quality of drinking water is vital for human health. Peak water constraints result in people not having access to safe water for basic personal hygiene. "Infectious waterborne diseases such as diarrhea, typhoid, and cholera are responsible for 80 percent of illnesses and deaths in the developing world, many of them children. One child dies every eight seconds from a waterborne disease; 15 million children a year."

Vital aquifers everywhere are becoming contaminated with toxins. Once an aquifer is contaminated, it is not likely that it can ever recover. Contaminants are more likely to cause chronic health effects. Water can be contaminated from pathogens such as bacteria, viruses, and parasites. Also, toxic organic chemicals can be a source of water contamination. Inorganic contaminants include toxic metals like arsenic, barium, chromium, lead, mercury, and silver. Nitrates are another source of inorganic contamination. Finally, leaching radioactive elements into the water supply can contaminate it.

Human conflicts over water

Some conflicts of the future may be fought over the availability, quality, and control of water. Water has also been used as a tool in conflicts or as a target during conflicts that start for other reasons. Water shortages may well result in water conflicts over this precious resource.

In West Africa and other places like Nepal, Bangladesh, India (such as the Ganges Delta), and Peru, major changes in the rivers generate a significant risk of violent conflict in coming years. Water management and control could play a part in future resource wars over scarce resources.

Solutions

Freshwater usage has great potential for better conservation and management as it is used inefficiently nearly everywhere, but until actual scarcity hits, people tend to take access to freshwater for granted.

Water conservation

There are several ways to reduce the use of water. For example, most irrigation systems waste water; typically, only between 35% and 50% of water withdrawn for irrigated agriculture ever reaches the crops. Most soaks into unlined canals, leaks out of pipes, or evaporates before reaching (or after being applied to) the fields. Swales and cisterns can be used to catch and store excess rainwater.

Water should be used more efficiently in industry, which should use a closed water cycle if possible. Also, industry should prevent polluting water so that it can be returned into the water cycle. Whenever possible, gray wastewater should be used to irrigate trees or lawns. Water drawn from aquifers should be recharged by treating the wastewater and returned to the aquifer.

Water can be conserved by not allowing freshwater to be used to irrigate luxuries such as golf courses. Luxury goods should not be produced in areas where freshwater has been depleted. For example, 1,500 liters of water is used on average for the manufacturing of a single computer and monitor.

WATER MANAGEMENT

Sustainable water management involves the scientific planning, developing, distribution and optimization of water resources under defined water polices and regulations. Examples of policies that improve water management include the use of technology for efficiency monitoring and use of water, innovative water prices and markets, irrigation efficiency techniques, and much more.

Experience shows that higher water prices lead to improvements in the efficiency of use—a classical argument in economics, pricing, and markets. For example, Clark County, Nevada, raised its water rates in 2008 to encourage conservation. Economists propose to encourage conservation by adopting a system of progressive pricing whereby the price per unit of water used would start out very small, and then rise substantially for each additional unit of water used. This tiered-rate approach has been used for many years in many places, and is becoming more widespread. A Freakonomics column in the New York Times similarly suggested that people would respond to higher water prices by using less of it, just as they respond to higher gasoline prices by using less of it. The Christian Science Monitor has also reported on arguments that higher water prices curb waste and consumption.

Conversely, certain subsidies can lead to inefficient use of water. Water subsidies often involve contentious policy issues that are political in nature. In 2004, the Environmental Working Group criticized the U.S. government for selling subsidized water to corporate farms for an average price of only $17 per acre foot (326,000 gallons). In his book The Ultimate Resource 2, Julian Simon claimed that there is a strong correlation between government corruption and lack of sufficient supplies of safe, clean water. Simon wrote, "there is complete agreement among water economists that all it takes to ensure an adequate supply for agriculture as well as for households in rich countries is that there be a rational structure of water law and market pricing. The problem is not too many people but rather defective laws and bureaucratic interventions; freeing up markets in water would eliminate just about all water problems forever... In poor water-short countries the problem with water supply—as with so many other matters—is lack of wealth to create systems to supply water efficiently enough. As these countries become richer, their water problems will become less difficult". This theoretical argument, however, ignores real-world conditions, including strong barriers to open water markets, the difficulty of moving water from one region to another, water rights laws that prevent redistribution based solely on economic value, inability of some populations to pay for water, and grossly imperfect information on water use. Actual experience with peak water constraints in some wealthy, but water-short countries and regions still suggests serious difficulties in reducing water challenges.

CLIMATE CHANGE

Extensive research has shown the direct links between water resources, the hydrologic cycle, and climatic change. As climate changes, there will be substantial impacts on water demands, precipitation patterns, storm frequency and intensity, snowfall and snowmelt dynamics, and more. Evidence from the IPCC to Working Group II, has shown climate change is already having a direct effect on animals, plants and water resources and systems. A 2007 report by the Intergovernmental Panel on Climate Change counted 75 million to 250 million people across Africa who could face water shortages by 2020. Crop yields could increase by 20% in East and Southeast Asia, but decrease by up to 30% in Central and South Asia. Agriculture fed by rainfall could drop by 50% in some African countries by 2020. A wide range of other impacts could affect peak water constraints.

Loss of biodiversity can be attributed largely to the appropriation of land for agroforestry and the effects of climate change. The 2008 IUCN Red List warns that long-term droughts and extreme weather puts additional stress on key habitats and, for example, lists 1,226 bird species as threatened with extinction, which is one-in-eight of all bird species.

BACKSTOP WATER SOURCES

The concept of a "backstop" resource is a resource that is sufficiently abundant and sustainable to replace non-renewable resources. Thus, solar and other renewable energy sources are considered "backstop" energy options for unsustainable fossil fuels. Similarly, Gleick and Palaniappan defined "backstop water sources" to be those resources that can replace unsustainable and non-renewable use of water, albeit typically at a higher cost. The classic backstop water source is desalination of seawater. If the rate of water production is not sufficient in one area, another "backstop" could be increased interbasin transfers, such as pipelines to carry freshwater from where it is abundant to an area where water is needed. Water can be imported into an area using water trucks.The most expensive and last resort measures of getting water to a community such as desalination, water transfers are called "backstop" water sources. Fog catchers are the most extreme of backstop methods.

To produce that fresh water, it can be obtained from ocean water through desalination. A January 17, 2008 article in the Wall Street Journal stated, "World-wide, 13,080 desalination plants produce more than 12 billion US gallons (45,000,000 m) of water a day, according to the International Desalination Association".Israel is now desalinizing water at a cost of US$0.53 per cubic meter.Singapore is desalinizing water for US$0.49 per cubic meter. After being desalinized at Jubail, Saudi Arabia, water is pumped 200 miles (320 km) inland though a pipeline to the capital city of Riyadh.

2

RAINWATER STORAGE METHOD

Storage is usually the most expensive component of a rainwater system and often determines the type of filtration and pumping system. We provide a wide range of storage solutions from a 75 gallon consumer tank to custom-designed modular systems capable of storing millions of gallons to run large commercial or multi-family structures or irrigation systems.

COMPARING RAINWATER STORAGE OPTIONS

Storage tanks, usually the most expensive component of the rainwater harvesting system, come in a wide variety of sizes and types. When deciding on the type of tank to use, the main factors to consider include where you live and your budget. When choosing the size of storage tank or cistern, consider several variables: rainwater supply (local precipitation), demand, projected length of dry spells without rain, catchment surface area, aesthetics, personal preference, and of course, your budget.

A myriad of variations on storage tanks and cisterns have been used over the centuries and in different geographical regions: earthenware cisterns in India, large pottery containers in Africa, above-ground vinyl-lined swimming pools in Hawaii, concrete or brick cisterns in the central United States, and, in Texas and Colorado, galvanized steel tanks and site-built stone-and-mortar cisterns.

Tanks can be above or below ground. Factors such as soil, outside temperature ranges, and cost should be used to determine whether a tank is placed above or below ground. Some tanks are suited for above-ground placement (i.e. vinyl-lined swimming pools), where others can be used both above and below ground (i.e. polyethylene). Some types of tanks are built to be buried (i.e. polyethylene tanks designed for burial).

Consequently, understanding all the information about the options available is critical to making a good decision about the type of tank to purchase, since it should prove to be something you live with for a long time. Below is a general overview of the various tank types to choose from and some of their characteristics.

Fiberglass

Fiberglass tanks are light-weight, reasonably priced, and long lasting. They are built in standard capacities from small 50 gallon barrels to much larger 15,000 gallon tanks and are available in both vertical cylinder and low-horizontal cylinder configurations. Smaller fiberglass tanks (i.e. under 1,000 gallons) are expensive for their size, so polyethylene might be preferred. Tanks for potable use need to have a USDA-approved food-grade resin lining and the tank should be opaque to inhibit algae growth. The durability of fiberglass tanks has been thoroughly tested and proven to last for years and can be easily repaired. The fittings on fiberglass tanks are an integral part of the tank, eliminating one common potential problem - leaking fittings.

Polyethylene

Polyethylene tanks are probably the most common type of tank being sold today and are readily available at most farm, ranch supply, and large landscape retailers. They vary greatly in size, shape, and color, and can be used above or below ground.

Most of the tanks stocked by farm and garden supply houses are usually for above-ground installations. For buried installation, specially designed and reinforced tanks are necessary to withstand soil expansion and contraction. Polyethylene tanks are comparatively inexpensive, lightweight, and long-lasting and are available in capacities from small 50 gallon barrels to large 10,000 gallon tanks. They are lighter in weight than other types of tanks, including fiberglass, and consequently, are cheaper and easier to transport.

Polyethylene tanks tend not retain paint well, so use pre-painted (i.e. pigmented) tanks manufactured with opaque plastic. Black and dark colored tanks will absorb heat and thus, should be shaded or buried. The fittings of these tanks are aftermarket modifications and are easy to plumb. However, the fittings are not always tight, and should be checked for leakage occasionally.

In-ground polyethylene

In-ground polyethylene tanks are more costly for two reasons: the cost of excavation and the cost of a more heavily reinforced tank.

The latter is required if the tank is to be buried more than two feet deep. Burying a tank in soil with high clay content is not recommended because of the expansion and contraction cycles of clay.

For below ground installation, the walls of poly tanks must be manufactured thicker, and sometimes, an interior bracing structure must be added.

Swimming Pools

Above ground swimming pools are commonly used in Hawaii for rainwater catchment; and although used throughout the islands, they are not a good rainwater catchment solution. A few of the issues related to these pools as a storage device are: 1) The pool liner typically only goes up to the lip of the pool and slips into the pool. 2) Liners are not food grade quality and may in fact be made to retard bacterial growth in the pool and can be toxic. 3) Large diameters of these pools are not easily covered and therefore it is difficult to keep out debris and animals.

However, swimming pools are very inexpensive, readily available and an easy to install solution. If this is the only option available then take the following precautions: 1) Make sure the liner is approved by the Food and Drug Administration (i.e. not the standard liner but one made for storing and holding water). 2) Plan to have a support system (i.e. a post in the middle of the pool) to hold up the cover so it does not sag into the water. 3) Make sure the cover can be firmly attached to the top of the pool to keep out unwanted guests and debris.

Wood

For personal aesthetic appeal, a wood tank are sometimes a desirable choice. Wood tanks, similar to water towers at railroad depots, were historically made of redwood. However, modern wood tanks are usually of pine, cedar, or cypress wrapped with steel tension cables. Wood tanks are lined with plastic to increase longevity. For potable use, a food-grade liner should be used.

Redwood, as a tank material, has a great reputation for being durable and attractive. It contains no resins and has high levels of tannin, a natural preservative resistant to insects and decay. It is also a good insulator, keeping the water cooler in the summer and protects it from freezing the winter.

These tanks are available in capacities from small 700 gallon tanks to very large 37,000 gallon tanks, and are usually site-built by skilled technicians. As with metal tanks, they can be dismantled and moved to a different location, if required. These tanks are for above-ground use and not for use in dry, hot climates.

Redwood is very attractive, but it is expensive and not readily available. Pine is commonly used and although it does not have some of the characteristics of redwood, it is readily available and less expensive.

Metal

As with wood, galvanized sheet metal tanks can also be an attractive option. They are available in sizes that range from small tanks of 150 gallons to medium-sized 2,500 gallon tanks, and are lightweight and easy to relocate if required.

Most tanks are corrugated galvanized steel dipped in hot zinc to improve corrosion resistance. These tanks should be lined with a food-grade liner, usually polyethylene or PVC, or coated on the inside with epoxy paint. The paint or liner will extend the life of the metal and, if being used for potable water, must be FDA and NSF approved for potability.

These tanks are for above-ground use. Old or recycled tanks may contain lead and should be avoided. Brass and bronze fittings should not be connected directly to the tank as they will cause corrosion. Additionally, care should be taken when cleaning these tanks, as a film develops naturally on the inside of the tank, which coats the tank and inhibits corrosion.

Concrete

The most versatile of tanks, concrete tanks can either be poured in place or prefabricated. They can be constructed above or below ground. They can be owner or contractor-built. Poured-in-place tanks can be very attractive and easily integrated into new construction. For example, the tank can be placed under a patio, or a basement. Concrete tanks, once poured, are considered permanent.

One unique advantage of poured concrete is that the concrete will over time decrease the corrosiveness of rainwater by leaching into the water. This advantage of concrete tanks results in a desirable taste imparted to the water by calcium in the concrete being dissolved in locations where there is slightly acidic rainwater. For potable systems, it is essential that the interior of the tank be plastered with a high-quality material approved for potable use.

Underground concrete tanks are prone to cracking and leaking, especially when in clay soil. Leaks can be easily repaired, although the tank may need to be drained to make the repair. If building the tank yourself, it is recommended to involve the expertise of a structural engineer to determine the size and spacing of reinforcing steel to match the structural loads of a poured-in-place concrete. A product that repairs leaks in concrete tanks, Xypex(TM), is now available and approved for potable use.

Ferrocement

Ferrocement is the term used to describe a steel and mortar composite material. These tanks can be above or below ground and can be done by contractors or homeowners. They were developed in third-world countries to be relatively low-cost and durable. They are listed separately from concrete, not just because of the materials used to construct them, but also because they have differing problems and advantages. These tanks are typically built with concrete, but have multiple layers of wire mesh - typically chicken wire-wrapped around a light framework of rebar, embedded in the concrete. Walls can be as thin as 1" and still be strong. Consequently, it can cost less to build than a concrete-only tank. If buying a ferrocement tank, make sure it does not contain any toxic compounds in the concrete and that the wires are not visible on the inside of the tank.

Ferrocement, like concrete, will need maintenance and repair as cracks appear. It is important to ensure that the ferrocement mix does not contain any toxic components. Some sources recommend painting above-ground tanks white to reflect the sun's rays, reduce evaporation, and keep the water cool.

Stone and Mason

Hand made stone or mason tanks are not as common as they once were. Increasing labor costs, decreasing costs and increasing availability of other types of tanks has limited their use to areas where labor is very, very cheap (i.e. third-world countries) or where budget is not an issue.

The mass of the stone gives these tanks two distinct advantages: it keeps the water cool in hot climates and they can be very attractive. As with ferrocement tanks, care should be taken to make sure the mix does not contain toxic materials.

These tanks are custom-built, so they can be as large as designed. Most tanks are designed to be circular, since this shape is more structurally sound. These tanks, if properly constructed and maintained, will last for decades.

Plastered Tire Cistern

Another type of hand-made tanks are plastered tire cisterns. They were pioneered with the earthships in the Taos, New Mexico area and some have been in use for decades. They are simply a circle of buried tires with a wire mesh inside covered with plaster. Just like stone or cement cisterns, they will need periodic maintenance to repair cracks on the inside.

These tanks are meant to be buried and can be very economical for large tank sizes (i.e., 10,000+ gallons/37,900 liters), especially if owner-built. In earthships, tanks are typically built as an integral part of the home and can provide cooling in hot climates. In cooler climates, the tops and sides should be insulated to prevent cooling. Care should be taken in building a plastered tie cistern to ensure the wire mesh and tires are thoroughly covered with plaster.

As with stone and mason cisterns, these tanks are custom-built, so they can be as large as designed. They are typically designed to be circular, but since they are buried, they can be almost any shape. These tanks, if properly constructed and maintained, will last for decades.

GENERAL GUIDELINES

General guidelines for all tanks include:

- Before you start, conserve, conserve, conserve. Cutting your water usage will reduce the size of tank you need and save you money.
- Remember - water is very heavy. (i.e., 500 gallons weighs over 2 tons!)
- Make sure the tank is easy to access and maintain.
- Tank should be opaque or darker, either upon purchase or painted later, to inhibit algae growth.
- For potable systems, storage tanks must never have been used to store toxic materials.
- Tanks must be covered and vents screened to discourage mosquito breeding.
- Tanks used for potable systems must be accessible for cleaning.
- Install first-flush and screening devices prior to water reaching the tanks to keep it as fresh and clean as possible.
- Keep tops of tanks free of debris to make it harder for animals to reach the top of the tank
- Buried tanks should be located in well-drained soil and location.
- Water weighs about 8 pounds per gallon so plan your pad, if any, before installing your tank.

- Plan where storage tank overflow should be piped or directed to. Keep it away from underneath your holding tank to prevent pad erosion and to keep animals away.

COLLECTING AND USING RAINWATER AT HOME

Rainwater harvesting collecting and storing rain for later use is an ancient practice. It is still used in many rural places throughout the world, and today it is making a comeback in urban centres as an additional source of water.

At its simplest, rainwater harvesting consists of a rain barrel placed under the downspout of your home to collect rainwater for garden irrigation.

Larger, more sophisticated, systems can be incorporated into your home's plumbing system to provide water for a variety of household needs, from toilet flushing to laundry, and even drinking water (in these instances the rainwater will be treated prior to use). All rainwater harvesting systems, simple or complex, have the same basic components:

A catchment area to capture the rainfall this is typically the roof of the house.

A conveyance system to move the water from the roof to a storage area eavestroughs and downspouts, and maybe piping.

A storage system to hold the rainwater for future use a barrel, a cistern or a tank.

A distribution system to get the water from storage to where it is being used this can range from a watering can to full integration with the existing plumbing system in the house.

Rainwater harvesting is advantageous for a number of reasons:

It can offset the amount of municipally treated water you use in your home helping to reduce your water bill

Larger cisterns can help divert and retain runoff from your property reducing the impact on local stormwater infrastructure and combined sewer systems

Stored rainwater can be used for landscape irrigation even during watering bans and helps to replenish groundwater supplies with water that would normally flow to stormwater systems

For those on well water, rainwater harvesting offers a way to reduce well pump energy consumption and reduce the amount of water the well has to deliver an important consideration if your well is affected by dry spells or water table fluctuations.

GETTING STARTED

When thinking about harvesting rainwater for your home, there are a number of questions to consider.

1. What would you like to use rainwater for?

Are you simply looking to harvest rainwater for outdoor watering purpose, or are you interested in a system that provides water for indoor uses as well (e.g., toilets, laundry)? Or are you wondering about using rainwater on a seasonal basis for a variety of purposes at the cottage? Each scenario involves different set-up, equipment, maintenance and costs.

2. Which rainwater uses are approved in your area?

The use of rainwater and other non-potable water supplies is regulated differently in each province and territory. The 2010 National Plumbing Code of Canada permits the use of non-potable water (including rainwater) for toilet and urinal flushing as well as outdoor irrigation. However, provincial, territorial and local jurisdictions may have different positions and requirements so it is prudent to check with local building authorities to see what's allowed in your specific area and what's not - before constructing your system.

3. How much rainwater do you need?

Depending on what you wish to use your rainwater for, the amount you need will vary depending on:

Household water consumption habits e.g., how often do you water outdoors and how much water is used, how many toilet flushes are there in a given period, how many loads of laundry?

The number of people in your household.

The age and efficiency of your plumbing fixtures (e.g., toilets/washing machines). If they are not water efficient, you may wish to replace them with some that are to make the most of both the rainwater you harvest and the municipal water you purchase.

CMHC's Collecting and Using Rainwater At Home, A Guide for Homeowners includes a worksheet on residential water consumption per household activity based on average consumption rates. This will help you calculate how much rainwater is needed to meet various requirements in your household on a daily, weekly, monthly and yearly basis depending on what is allowed by the building authority in your area.

4. How much rainwater can you collect?

The amount of rain you can harvest is calculated based on the square footage of your catchment area and the climatic rainfall conditions in your area. The general rule of thumb is that about 80% of the annual rain that falls on a catchment area can be collected; the rest is lost due to overflows, leaks, absorption by the roofing materials, and so on. Refer to the guide (see above) on how to determine how much rainwater you may be able to collect over the course of the year.

5. How much rainwater can you store?

Rainwater storage tanks are available in many sizes, from less than one hundred litres suitable for supplementary outdoor plant watering to thousands of litres for larger residential systems.

The Rainwater Guide offers help with determining optimal tank size for a specific volume of water needed based on your roof area and local climatic conditions.

DESIGNING YOUR SYSTEM

This section provides an overview of the key components and features of a rainwater harvesting system.

Roof Catchment Area

Certain roofing materials can impact rainwater quantity and quality. For instance, porous materials such as asbestos shingles, clay or concrete tiles, or wood shingles will soak up some of the rainwater, which reduces the amount harvested during a rainfall. Other non-absorptive roofing materials, such as steel, shed water readily for collection. Certain roofing materials may contain potential contaminants (heavy metals, and pesticides) that can be absorbed by the rainwater as it flows over the roof surface, making it unsuitable as drinking water, or for use on edible plants.

Eavestroughs and Downspouts

To get the full benefit of the catchment area, your home needs to be equipped with properly sized and sloped eavestroughs (gutters) and an adequate number of downspouts. Otherwise water may overflow during heavy rainstorms and get absorbed in the ground or run off into the storm sewers. It may be necessary to have an in-ground collection system that gathers all the water drained by downspouts located at different positions around the house to take full advantage of the rainwater that falls on the roof.

RAINWATER CONVEYANCE NETWORK PIPING

The conveyance network includes the plumbing used to take the rainwater to the storage tank and to distribute it from the tank to the various end-use points. Pipe materials, sizes and slopes are regulated by local and provincial codes. Above-ground piping should be UV-resistant and structurally strong enough to keep it from deteriorating or being damaged.

Rainwater Storage Tanks

Tanks can be located above or below ground, outside or inside the home, garage or shed, and are available in many different sizes and materials.

Polyethylene or polypropylene tanks are the most common. They are durable, lightweight and available in a range of sizes, shapes and colours, and typically suitable for both above- and below-ground installation. Concrete tanks are generally only used in below-ground systems. They can be cast in place as part of a building's foundation or installed separately elsewhere on your property.

Other types of materials include fiberglass (above or below ground), galvanized steel (low cost, above ground only), stainless steel (more costly, above ground only), and wood (aesthetically pleasing, above ground only).

Protective Measures

A well-designed system includes an overflow pipe to protect against damage caused by the tank overflowing during periods of heavy rains or low usage. All overflow water should be discharged away from foundations and other structures.

Tanks, conveyance networks and distribution systems must be protected against freezing. All below-ground components should be installed below the frost line. Seasonal systems should be completely drained at the end of the season.

In plumbing systems where there is a possibility that the rainwater conveyance piping could become interconnected with the household potable water piping, the use of backflow prevention devices ensures that non-potable water cannot flow backwards into the potable water system and contaminate it this is required by the 2010 National Plumbing Code of Canada.

Rainwater Treatment and Filtration Components

Rainwater can become contaminated from leaf litter and debris on your roof as well as from dust and local air pollution. Care should be taken to remove contaminants before they collect in your storage tank. Pre-filtration devices can be incorporated in the conveyance system to remove contaminants that may collect on the roof, before the water reaches the storage tank. Options include gutter guards, downspout filters and "first-flush" devices to divert the first several litres of rainwater from the roof away from the conveyance system at the start of each rainfall.

Post-storage Treatment

Depending on the intended use of your rainwater, you may wish to consider some form of filtration and/or disinfection of the water that comes from the storage tank to reduce potential health risks, particularly if you wish to use the harvester rainwater for drinking. Check with your local health and building departments.

Preparing for Installation

Proper installation of a rainwater harvesting system is essential. While a simple rain barrel can be placed almost anywhere with very few restrictions (apart from ensuring it does not spill water against the wall or foundation of the house), a large system must be installed safely, in accordance with local and provincial codes and standards. Whether you choose an above-ground, indoor or underground tank, there are a number of issues to consider.

Site Selection

Proper site selection will help prevent the rainwater system from causing damage to, being damaged by, or interfering with existing property features. Factors include location of utility services, wells and septic tanks; topography, depth to bedrock and soil classification; property easements and rights-of-ways, and visibility and aesthetics for you and your neighbours.

Tank Placement

Water is heavy 1,000 litres (enough to fill a 1 metre by 1 metre by 1 metre cube) weighs 1,000 kg so the tank will need adequate structural support. Depending on local regulations, above-ground tanks may be placed on a flat, level concrete or compacted earthen or gravel base. In some municipalities, the base must be designed by a structural engineer. Whether indoors or outdoors, the tank also needs to be secured to prevent it from falling over. Anchoring devices or ballasts are required for below-ground tanks to ensure that buoyancy doesn't push them to the surface.

Health and Safety

The health and safety implications of your rainwater harvesting system need to be addressed. All plumbing and electrical work must be undertaken by certified plumbers and electricians, in accordance with local codes and regulations.

Further, the 2010 National Plumbing Code of Canada requires that plumbing that provides non-potable water within a residence must be labeled accordingly. This will prevent plumbers or Do-It-Yourselfers from accidentally connecting non-potable supplies to the potable supply and thereby possibly contaminating the drinking water and placing the household at risk. Check your local regulations for precautions that must be taken in your area.

Maintenance

A rainwater harvesting system needs ongoing monitoring and regular maintenance to protect against blockage, contamination and unwanted leaks, and to ensure optimal performance and water quality.

Most of the maintenance routines are quick and easy to perform, and require mainly a visual inspection of the system components every 3,6 and 12 months, with replacements and repairs as needed. Some municipalities may require an annual mandatory inspection by a licensed plumber. It is also recommended that the storage tank be drained every 3 to 5 years to check for deterioration and clean out accumulated debris. CMHC's Collecting and Using Rainwater At Home, A Guide for Homeowners, contains a detailed inspection and maintenance schedule as well as troubleshooting guidance.

COSTING YOUR SYSTEM

Because each system is unique, its cost is unique as well. A rainwater harvesting system can cost anywhere from a few hundred to thousands of dollars, depending on its size and level of sophistication. Costs may also vary by region and by availability of trades and rainwater harvesting experts.

Get Estimates

Talk with several suppliers and/or rainwater harvesting experts and get quotes, so you can compare and can choose based on the best value for your investment. Use the costing worksheet in CMHC's Collecting and Using Rainwater At Home, A Guide for Homeowners to make sure you are including everything needed for your system and to keep track of quotes.

After-Storage Treatment

Treatment of the water after the storage and before use is critical for both health of the users and maintenance of the system. The level of treatment will depend of the intended use of the water. For example, water used for landscape drip irrigation would not need the same level of treatment as water used for potable indoor purposes. In any case, we recommend a multi-barrier approach where more than one method of treatment are used to maximize effectiveness. For example, a common treatment-train for potable water in rural Texas is a set of cartridge filters followed by a ultraviolet light. Below are the different types of water treatment available.

FILTRATION

Filtration is similar to screening but on a smaller scale. There are various levels of filtration, and they are measured on a "micron" level. This measures the diameter of a particle that would be blocked by the filter. For example, a 5-micron filter would block particles in the water that are 5 microns or larger. (a micron is one-millionth of a meter). Filters can remove microorganisms, sediment, metals, and other organic matter. If larger sized filters are used, small microorganisms, such as bacteria can pass through, so a disinfection method is needed (explained next). It is important that they are checked and changed on a regular basis so that they maintain their effectiveness.

The goal of disinfection is to destroy the microbiological organisms that have the potential to cause illness or harm. Some of the smaller microorganisms, such as bacteria, may pass through large cartridge filters so they must be targeted with disinfection. Three disinfection methods common to RWH systems are chlorination, ultraviolet light (UV), and ozonation.

Chlorination uses the chemical of chlorine in either dry, liquid, or gas to kill microorganisms. It is very effective with viruses and bacteria. An advantage to chlorine is that is leaves a residual so that your entire distribution system remains disinfected. However, using chlorine can be dangerous, so it's important to follow manufacturer's guidelines. Only use chlorine that is approved for potable water use (standard ANSI/NSF 60). For more information see this guidebook by the TCEQ.

UV light is a common disinfection method used by Texas residents. The device works by sending water through a tube that contains the UV bulb (see picture on left). The light disrupts the DNA of microorganisms so that they can't reproduce. There are a Class A and Class B UV light. The Class A light is rated to destroy pathogens and is required to have an alarm to alert users if the device is not working. A Class B light cannot be used as a sole method of disinfection but helps to polish water that has already been treated. An important point to keep in mind is that UV lights are most effective when the water is clear. Any sediment in the water can block the pathogens from the light. This is why have the filtration BEFORE the UV light is so critical. The UV light is harmful to look at directly, so use approved safety glasses when maintaining the device. Ozonation disinfects by introducing ozone gas to the water. It is usually done at the point where water is used in the distribution system or in the storage tank. It is a colorless gas that disinfects, oxidizes, deodorizes, and decolorizes. Ozone gas is toxic and installation and maintenance of this type of system must be done by a licensed professional.

Adsorption uses activated carbon (charcoal filters) to bind to harmful contaminants, such as VOCs. These are the types of filters commonly used on faucets. They can also be used to remove chlorine from water used for disinfection.

When I was growing up, I remember drinking out of a rain barrel with a ladle. My great aunt would yell out the door, "Remember not to drink off the top!" That was then and now is now.

A lot has changed in 4 decades. There are a lot more pollutants, and we are more aware of the risks. We now know that E. coli and other harmful bacteria can be passed along in untreated contaminated water. A report by Peter H. Gleick estimates that if no action is taken to address unmet basic human needs for water, as many as 135 million people will die from water-related diseases by 2020.

Rainwater harvesting is viewed by many, including the EPA, as a partial solution to the problems posed by water scarcity: droughts and desertification, erosion from runoff, over-reliance on depleted aquifers, and the costs of new irrigation, diversion, and water treatment facilities.

Harvested rainwater in the U.S. is used mostly for irrigation; however, there is a growing interest in using rainwater for drinking and other indoor uses. Over 50% of household water is used indoors; bringing rain indoors could save the expense and environmental costs of treating and transporting water.

Can rainwater be made safe to drink? Yes. How safe? As safe as your well or tap water. How do you make it safe for indoor use? By filtering and purifying it.

Contaminants in water may include algae, air pollution, bird excrement, and leaves, sand, and dust. Local wells have dealt with these problems for decades. Installation of filtration and purification equipment can remove these contaminants at home as well.

First, take measures to keep foreign matter out of the incoming rainwater. First flush devices, gutter screens and other screening mechanisms keep the rainwater as clean as possible before it enters the conveyance system. Using screens and filters will greatly reduce maintenance and lengthen the life of the pump and filtration/purification system.

Even the best screening systems will allow unwanted particulates into the cistern. To keep sediment where it belongs, at the bottom of your tank, screen incoming rainwater, give the remaining sediment time to settle, avoid disturbing it, and don't pull water from the bottom of the tank. Use a floating filter, which extracts water from the middle of the tank, leaving sediment undisturbed.

Next is filtration, which removes debris from the water. Disinfection or purification follows, which kills contaminants and removes harmful substances that may be present. To determine what type of system you need, test the rainwater at a reliable laboratory. Without testing, you could spend a lot of money on equipment that will not give you safe water.

Filtration is included in every system, even simple irrigation systems. Examples of filtration systems include: screen filters, paper filters, and carbon or charcoal filters. Almost all systems use multiple filters. For example, after gutter screens and/or a first flush device, a system often includes two in-line filters of increasing fineness, a carbon filter and a UV light. Each of these are described below to assist you in evaluating what might be the right alternative for your planned water use and required water quality.

In starting to evaluate filter options, it is imperative to know exactly what the filter system you select will actually remove from the water. National Sanitation Foundation/American National Standards Institutions (NSF/ANSI) standards are the best, most stringent in the industry. Almost all water-filtration products are certified under NSF Standard 61 for Drinking Water System Components (see Related Topics). But the critical standards for contaminant removal are Standard 42, "Drinking Water Treatment Units – Aesthetic Effects," and Standard 53, "Drinking Water Treatment Units – Health Effects."

Standard 42 covers specific aesthetic contaminants (chlorine taste and odor, and visible particulates). Standard 53 covers health-related contaminants, such as Cryptosporidium, Giardia, lead, and volatile organic chemicals that may be present in drinking water. Systems that meet both of these standards are available, but expensive. Fortunately, the NSF website (see Related Topics) provides an easy way to search for units made by a specific manufacturer or that remove a specific contaminant.

FILTERS AND DISINFECTION

Filters are measured in microns. One micron is about 1/25,000th of an inch. For comparison, sand is about 100 – 1,000 microns, a human hair is about 100 microns, a particle of dust is about 1 micron and a virus can be smaller than .01

micron. The first filters in a system are cartridge filters. They range widely in what they are capable of removing and are used in a series (e.g., a 20 micron followed immediately by a 5 micron filter).

Filters are rated by the smallest size of particle they are capable of filtering. The smaller the micron size the better the filter. However, the finer the filter, the higher its cost and the slower its process. Filters have to be changed regularly, as an old, used filter is an excellent environment for microorganisms and potentially harmful pathogens.

For wells and rainwater systems a larger (e.g., a 50 micron) filter or equivalent screen (e.g., 300 mesh) should be used first to eliminate sand and large particles. This screen should be easily accessible and cleaned quarterly. Next is a 20 or 10 micron filter, followed immediately by a 10 or 5 micron filter. These are cleaned less frequently, but at least annually.

Filters will not eliminate all substances in the water. To create drinking quality water, filtration is always followed by disinfection. The EPA requires surface and ground water to be disinfected before it is consumed. Consequently, public water systems add disinfectants to destroy microorganisms that can cause disease in people and animals.

This is also necessary for rainwater, as the natural environment contains many microorganisms. Most are not harmful to us. Some, however, such as Giardia lamblia, can be deadly. These need to be eliminated from water before it is consumed.

Kinds of disinfection include chlorinization, ozonization, ultraviolet (UV) light, and membrane filtration. In evaluating disinfection methods, be aware that some actually create unhealthy byproducts that need to be treated.

The effectiveness of disinfection is judged by looking for an indicator organism that, if present, indicates other more harmful pathogens may be present. In getting a water test, this indicator organism is Total Coliform Bacteria that, if present, indicates other pathogens may be present as well.

Chlorine has been used as a disinfectant in public water systems for most of the past century. The introduction of chlorine to disinfect water has virtually eliminated waterborne diseases such as cholera, typhoid, dysentery and hepatitis, saving thousands of lives. However, it is often maligned due to suspected side effects.

For disinfection purposes, 2.3 fluid ounces of household bleach must be added per 1,000 gallons of water. Chlorine dosage rate will vary depending on quantity of water to be treated, pH and temperature. A major downside of chlorine is that it is very reactive and easily combines with naturally occurring organic material to create harmful trihalomethanes (THMs) like chloroform. Chloroform is formed when chlorine reacts with either humic and/or fulvic acids, which are commonly found in water.

Because chlorine is reactive, it quickly dissipates. Keeping the dosage rate correct is critical when using this method of disinfection. THMs should be tested for in the water source if you are going to use Chlorine.

To reduce the possibility of harmful byproducts with the use of Chlorine, do the following:

Remove the byproducts after they have been created. This is costly, typically meaning other purification systems must be employed (e.g., Reverse Osmosis or other purifcation systems) or The concentration of particulates/organics in the water before it is treated. This is accomplished by using filters to remove these substances from the water prior to chlorine treatment.

The Chlorine smell and taste can be removed with an activated carbon filter, often referred to as a charcoal filter. Granulated activated carbon filters are sometimes made from coconut shells and can be considered a "green" solution. Carbon block filters are compressed activated carbon, fused with a binding substance into a solid block.

ULTRAVIOLET LIGHT

An alternative for disinfecting water is Ultraviolet (UV) light. UV lights have been used for nearly a century in Europe and are now common in the US. With UV lights, the water must always pass through a filtration system first. If no filter is used, pathogens and bacteria will cast shadows in the flowing water, thereby allowing live organisms to pass through unharmed.

UV light works by penetrating an organism's cell walls and disrupting the cell's genetic makeup, making it impossible to reproduce and rendering it harmless. Often it is claimed that it kills the microorganism, but it doesn't - it just makes them unable to reproduce and thus harmless. UV lights do not change the chemical composition of the water and leave behind no by-products.

For UV to be effective the right light dose must be used to a specific unit of water and the water must be clear of suspended solids and other particulates. Most UV units are usually insensitive to temperature and pH differences in the water, but manufacturers' fine print should be read and followed.

There are several issues with UV lights should be taken into consideration:

Replace the bulb at the manufacturer's specified intervals – generally after 9,000 hours, or about every 12 months;

UV light is not visible to the human eye, so it may appear to be lit and in fact is not working;

The glass enclosure around the light needs to be cleaned occasionally for the UV light to be effective;

If no backup light is installed the water needs to be shut off upstream of the bulb prior to replacing the unit. Generally it is prudent to disinfect the water downstream after the system has been shut down for any reason.

UV system Correct UV treatment is effective in reducing harmful pathogens from the water. It is generally recommended that home units include alarms to notify the user when a bulb needs to be serviced or the unit is not working. Purchase a unit that has an automatic bulb cleaner, to reduce maintenance requirements. Two units should be installed, so when one unit needs servicing the second unit can be turned on so there is no disruption in disinfecting the drinking water.

UV light manufacturers rate their systems to a given dosage at a given flow rate (e.g., 10 gallons per minute). When installing a UV light, make sure the flow rate of the UV unit is matched to your flow rate of water (i.e., the pump flow rate). If the pump rate is greater than that of the UV light, install a pressure regulator or flow restrictor.

To properly treat the rainwater, it must contain particulates no larger than 50 microns and contain no tannins, sulfur or sulfur-related bacteria, have less than 0.3 parts per million of iron, and less than 0.005 parts per million of manganese. Knowing whether these are in the water and need to be treated is a great reason to test your water before installing a system. If any of the above is present in the water, the filters must deal with these elements before the water is treated by a UV light. Most of these will not be present in rainwater, but could result from local air pollution or contamination of the conveyance system. Don't assume anything until your water has been lab tested.

The UV light unit is typically installed after all filtration and the resulting water is clean, bug-free and ready to use. Entry-level units will handle about 10 gallons per minute. The price of the unit will increase as options and flow rates increase.

MEMBRANE FILTRATION

Membrane filtration is another alternative. Membrane filtration involves pushing water through a layer of material. Pressure-driven membrane technologies include microfiltration, ultrafiltration, nanofiltration and reverse osmosis. It is one of the few technologies capable of removing pharmaceuticals, and creates no byproducts.

Membrane technologies are more costly than other alternatives, but prices are rapidly declining. Most water purification experts expect membrane technology to become the prevalent technology in smaller systems over time as their price drops.

Choosing the right membrane technology is not straightforward, as the technology is changing and there are no real standards. Make sure you know what you need and match it to the type of system you are evaluating. Again, it is critical to test your water to know what you need before evaluating options.

Microfiltration (MF) is a membrane separation process using a pore size of .03 to 10 microns. Although this does not sound like a big range, when it comes to water purification, it is. The smaller the pore size, the more the system will remove. Microfiltration membranes are good for the removal of sand, silt, clay, algae, cysts and some bacteria.

Ultrafiltration (UF) is a membrane separation process using a pore size of approximately .002 to .1 microns. UF will remove all materials removed by an MF system, plus some viruses.

Nanofiltration membranes (NF) have an approximate pore size of only .001 microns. These small pore sizes require much more power to push water through the membrane and generate more waste than either MF or UF filtration systems. These systems eliminate virtually all cysts, bacteria, viruses, and other materials, including minerals. Consequently, the resulting water has a low pH that can be corrosive and needs to be remineralized, commonly using limestone, to raise the pH. Due to the greater power requirements, NF has yet to become mainstream.

Reverse Osmosis (RO) is the most widely used membrane technology today. Under counter RO systemThese systems remove particles as fine as .001 microns, are compact, simple to operate and have been in use for over a decade. RO systems remove radium, natural organics, pesticides, cysts, bacteria and viruses. To ensure contaminant reduction, seek out units certified by NSF for contaminant reduction and not just safety. RO systems produce waste water that needs to be processed; however, the newer units are becoming "greener," producing less, but still significant, waste. These units vary greatly in their efficiency, so make sure to ask about waste and efficiency when shopping for an RO system.

RO waste water contains a high concentration of the contaminants removed from the water, so dealing with this waste must be planned for when installing an RO system. Options for dealing with this water include plumbing through a greywater system to the irrigation system or directly to the septic system.

RO systems come in small under-the-counter units or whole-house systems. Prices will vary greatly for these units and only NSF-certified units should be considered. Under-the-counter units generally include a sediment filter, a carbon filter, the RO membrane and another carbon filter, and will generally cost under $1,000. A whole-house unit contains all the same components, but is capable of handling much larger water flow rates, and generally includes a calcite or equivalent filter to reduce the pH of the water, and a large storage tank (e.g., 20 – 50 gallons). The cost of a whole-house unit can run upwards of $8,000, depending on size of the house and family.

Regardless of system size, maintenance needs to be performed regularly. The most frequent maintenance is changing cartridges. Filters are used to protect the RO membrane from particle fouling. As these filters trap particles from the water supply, a reduction in pressure occurs. Many RO units include a low-pressure switch that prevents the RO from running if the pressure drops too low. Check the allowable pressure drop across the cartridge and compare this to the incoming feed pressure. If it is lower than manufacturer recommendations, the filters need to be replaced.

Distillation

The last commonly available purification technology is distillation. Distillation separates the water from the impurities through heating and then collecting the condensation. It is very energy intensive and loses about 5-10% of the water due to evaporation. Distillation removes almost all substances from the water with the exception of volatile organic chemicals (VOCs) that evaporate easily. To this end, some distillation systems are also equipped with carbon filters to remove the VOCs.

Distillation works slowly to reduce energy requirements and, like RO systems, will store the purified water in a tank for later use. In addition to using a lot of electricity to operate, distillation systems generate heat.

Distillation units producing 5 -12 gallons of water a day will typically cost about $1,500 - $2,000. Cost will increase as capacity increases and as options are added. High-end automatic home units with larger storage capacity may cost upwards of $4,000. New solar distillers give you the option of reducing the electrical requirements.

Standard Practice for Household Use

A common practice in off the grid homes is to filter all the incoming rainwater and then store it in a small pressure tank. From the pressure tank the outgoing water is split into two separate paths - one path for potable and the other for non-potable water. A purification process is added to produce potable water. The major advantage of this approach is that it requires a much smaller unit and costs less, since it treats less water than a whole-house unit. But the disadvantage is that it requires a dual plumbing system – one to supply filtered but non-potable water to the toilets, clothes washer, irrigation faucets, etc., and one to supply potable water to the faucets.

An apparently low-cost, entry-level system is a countertop or pitcher type unit for potable water. However, when measured on gallons of water processed between changing filters, these units tend to be much more expensive in the long run. For example, a typical faucet unit available at most large hardware stores needs its filter changed every 100 gallons. For a family, this would be more than once a month and each filter costs about $30. This could cost nearly $500 a year, just for filters!

Before investing in filtration or purification equipment, invest in removing particulates before they enter into the system by installing gutter screens, leaf screens and roof washers. Removing materials before they enter the system is far easier and less expensive than dealing with them afterwards.

There is no perfect solution for disinfecting water, as all solutions have some environmental cost. Some require substantial energy, some create harmful by-products and some waste water. To save money, test your water (have you heard that before?) and get the right unit to solve your specific problem. Generally, the smaller the capacity the less expensive the unit will be overall, so get only what you need.

Lastly, remember that as the owner of a water system, it is your responsibility to maintain it. When you pay for utility-purified and -delivered water, maintenance is included in your bill. But when you own your water system, it is your responsibility to maintain it on a regular basis.

Rainwater can be safely used outdoors and indoors if the correct steps are taken to handle, store and clean it. Although not yet common in the US, indoor use of rainwater is practiced worldwide. As population growth continues, water rates increase and the desire to be "more green" and self-reliant increases, rainwater use will become more common here in the United States.

The Adoption of Water Conservation Application Techniques

A number of farm application irrigation technologies that could contribute to the conservation of water in both surface and piped systems have been developed. These techniques are generally adopted for the cultivation of high value crops. About 20 million hectares worldwide are now irrigated by sprinkler irrigation. Micro-irrigation as a technology has matured into reliable water and fertigation management system for crop production over the past decades and its usage continues to increase rapidly. A total of 3.2 million hectares are being irrigated by micro-irrigation techniques throughout the world, which represent one percent of the total areas irrigated in the world. A growth of more than six times over the last 20 years has been achieved, which is quite a success. Micro-irrigation is an important tool in the drive towards food security. Micro-irrigation systems have great potential to create significant opportunities for smallholder agriculture.

During the 1980s micro-irrigation started penetrating developing countries, mainly India. A few years later, fearing a water and food shortage, the Chinese Government declared drip irrigation technology to be amongst the top priorities for development in the last decade. However, Vidal et al., (2001) have shown that micro-irrigation "is not a miracle technology: excellent as well as poor results were obtained from these technologies, and their adoption really depends on farmers capacity to finance and operate them, as well as on the type of crop production".

According to the ICID database on modern irrigation techniques, it appears that the use of pressurized water application techniques is widely different between countries. The percentage of modern technique decreases with the total area irrigated in developed countries from a low of 21 percent in the USA and 16 percent in Italy to 90 percent or over in Northern European countries (France and Germany). Sprinkler has a much greater application compared to that of the drip systems with the exception of Cyprus, Israel and Jordan.

In South Asia and Africa low cost bucket-and-drip sets are becoming increasingly popular with farmers. In areas where shallow groundwater is plentiful, thousands of poor farmers in Bangladesh have used low-cost treadle pumps to supply water for crops for their own food security and additional income.

Advanced surface irrigation techniques such as precise land levelling, gated pipes or cablegation have also been developed to improve the typical poorly distributed application of water under conventional methods by furrow, border or basin irrigation. Experience shows that these techniques are too sophisticated to be adopted by most smallholders in developing countries. Adaptive research in the Middle East and North Africa has shown the difficulties in introducing surface water conservation techniques. Many attempts to introduce gated pipes in Morocco for example have failed for a number of reasons, mainly the poor maintenance of the quaternary earth canals. Considerable efforts for training of farmers in water conservation in surface systems are needed.

Farmers have adopted pressurised systems much faster than surface application techniques. For example, about 65 percent of the farmers in the Jordan valley have shifted from surface to drip irrigation over a 10-year period. Farmers built on-farm storage reservoirs to provide the flexibility required for drip irrigation.

Although there are no statistical data available on it, it is likely that most of the areas irrigated through water conservation techniques are served from wells or piped systems. There is still a huge potential for development of water conservation techniques in the areas presently irrigated from groundwater. Application of these techniques will continue to expand with the market for high value crops. However, the adoption of these techniques is limited to the cultivation of certain crops because of technical and financial considerations.

Canal Irrigation Systems: Water Efficiency and Water Productivity

Much more challenging than that of groundwater is the improvement of performance of canal systems, which will remain the dominant water transfer technique in developing countries in the future. The standards used for design of irrigation canals in many countries have not kept up with the development of new technologies. Most traditional delivery systems have no or little flexibility built into them. Their control strategies and control structures are inadequate for good water management practices. Water deliveries at farm level are not compatible with the key requirements of modern agriculture. Thoughts on how to assess the performance of canal projects are evolving.

The engineering concept of efficiency at different levels of a canal system is important for planning, designing and even operating a canal irrigation system. However, water resources experts have criticised this concept which could be misleading. These experts argue that seepage losses contribute to the recharge of aquifers and losses from a project can be recovered downstream if the water quality has not deteriorated beyond certain limits. These arguments are valid when considering the development of water resources at basin level or even at broader levels when aquifers extend beyond basin boundaries. Advocates of water conservation measures, such as canal lining, then respond that water quality may decline in the recycling process, lands adjacent to canals may be affected by waterlogging and salinization, and an important volume of water can be lost from wasteland. Furthermore, farmers have to support the cost of groundwater pumping to which opponents can argue that farmers use groundwater more efficiently than canal water which is too often unreliable. The application of water conservation measures in irrigation can have negative or positive impact on third parties. The theoretical debate is endless.

Efficiency alone is not a sufficient indicator to define the performance of an irrigation system. Indicators of water productivity expressed in terms of production or value of production per unit of water are very important in increasingly water-scarce situation. A canal irrigation system may have high conveyance efficiency with a minimum of seepage and operational losses. However, if water delivery is too rigid or unreliable, there will be considerable waste further down at the farm level. A water productivity indicator provides a global indication of the effectiveness of water conservation measures and of the quality of service provided to the users, as well as the farm use of water and other inputs.

25 ways to Conserve Water in the Home and Yard

Water conservation has become an essential practice in all regions, even in areas where water seems abundant.

In addition to saving money on your utility bill, water conservation helps prevent water pollution in nearby lakes, rivers and local watersheds.

Conserving water can also extend the life of your septic system by reducing soil saturation, and reducing any pollution due to leaks. Overloading municipal sewer systems can also cause untreated sewage to flow to lakes and rivers. The smaller the amount of water flowing through these systems, the lower the likelihood of pollution. In some communities, costly sewage system expansion has been avoided by community wide household water conservation.

Water conservation in the home...

1. Check faucets and pipes for leaks: A small drip from a worn faucet washer can waste 20 gallons of water per day. Larger leaks can waste hundreds of gallons.

2. Don't use the toilet as an ashtray or wastebasket: Every time you flush a cigarette butt, facial tissue or other small bit of trash, five to seven gallons of water is wasted.
3. Check your toilets for leaks: Put a little food coloring in your toilet tank. If, without flushing, the color begins to appear in the bowl within 30 minutes, you have a leak that should be repaired immediately. Most replacement parts are inexpensive and easy to install.
4. Use your water meter to check for hidden water leaks: Read the house water meter before and after a two-hour period when no water is being used. If the meter does not read exactly the same, there is a leak.
5. Install water-saving shower heads and low-flow faucet aerators: Inexpensive water-saving low-flow shower heads or restrictors are easy for the homeowner to install. Also, long, hot showers can use five to ten gallons every unneeded minute. Limit your showers to the time it takes to soap up, wash down and rinse off. "Low-flow" means it uses less than 2.5 gallons per minute.

   You can easily install a ShowerStart showerhead, or add a ShowerStart converter to existing showerheads, which automatically pauses a running shower once it gets warm. Also, all household faucets should be fit with aerators. This single best home water conservation method is also the cheapest!
6. Put plastic bottles or float booster in your toilet tank: To cut down on water waste, put an inch or two of sand or pebbles inside each of two plastic bottles to weigh them down. Fill the bottles with water, screw the lids on, and put them in your toilet tank, safely away from the operating mechanisms. Or, buy an inexpensive tank bank or float booster. This may save ten or more gallons of water per day. Be sure at least 3 gallons of water remain in the tank so it will flush properly. If there is not enough water to get a proper flush, users will hold the lever down too long or do multiple flushes to get rid of waste. Two flushings at 1.4 gallons is worse than a single 2.0 gallon flush. A better suggestion would be to buy an adjustable toilet flapperthat allow for adjustment of their per flush use. Then the user can adjust the flush rate to the minimum per flush setting that achieves a single good flush each time. For new installations, consider buying "low flush" toilets, which use 1 to 2 gallons per flush instead of the usual 3 to 5 gallons. Replacing an 18 liter per flush toilet with an ultra-low volume

(ULV) 6 liter flush model represents a 70% savings in water flushed and will cut indoor water use by about 30%.

7. Insulate your water pipes: It's easy and inexpensive to insulate your water pipes with pre-slit foam pipe insulation. You'll get hot water faster plus avoid wasting water while it heats up.

8. Take shorter showers: One way to cut down on water use is to turn off the shower after soaping up, then turn it back on to rinse. A four-minute shower uses approximately 20 to 40 gallons of water.

9. Turn off the water after you wet your toothbrush: There is no need to keep the water running while brushing your teeth. Just wet your brush and fill a glass for mouth rinsing.

10. Rinse your razor in the sink: Fill the sink with a few inches of warm water. This will rinse your razor just as well as running water, with far less waste of water.

11. Use your dishwasher and clothes washer for only full loads: Automatic dishwashers and clothes washers should be fully loaded for optimum water conservation. Most makers of dishwashing soap recomend not pre-rinsing dishes which is a big water savings.

    With clothes washers, avoid the permanent press cycle, which uses an added 20 liters (5 gallons) for the extra rinse. For partial loads, adjust water levels to match the size of the load. Replace old clothes washers. New Energy Star rated washers use 35 - 50% less water and 50% less energy per load. If you're in the market for a new clothes washer, consider buying a water-saving frontload washer.

12. Minimize use of kitchen sink garbage disposal units: In-sink 'garburators' require lots of water to operate properly, and also add considerably to the volume of solids in a septic tank which can lead to maintenance problems. Start a compost pile as an alternate method of disposing food waste.

13. When washing dishes by hand, don't leave the water running for rinsing: If your have a double-basin, fill one with soapy water and one with rinse water. If you have a single-basin sink, gather washed dishes in a dish rack and rinse them with a spray device or a panful of hot water. Dual-swivel aerators are available to make this easier. If using a dishwasher, there is usually no need to pre-rinse the dishes.

14. Don't let the faucet run while you clean vegetables: Just rinse them in a stoppered sink or a pan of clean water. Use a dual-setting aerator.

15. Keep a bottle of drinking water in the fridge: Running tap water to cool it off for drinking water is wasteful. Store drinking water in the fridge in a safe drinking bottle. If you are filling water bottles to bring along on outdoor hikes, consider buying a LifeStraw personal water filter which enables users to drink water safely from rivers or lakes or any available body of water.

    Water conservation in the yard and garden...

16. Plant drought-resistant lawns, shrubs and plants: If you are planting a new lawn, or overseeding an existing lawn, use drought-resistant grasses such as the new "Eco-Lawn".

    Many beautiful shrubs and plants thrive with far less watering than other species. Replace herbaceous perennial borders with native plants. Native plants will use less water and be more resistant to local plant diseases. Consider applying the principles of xeriscape for a low-maintenance, drought resistant yard.

    Plant slopes with plants that will retain water and help reduce runoff.

    Group plants according to their watering needs.

17. Put a layer of mulch around trees and plants: Mulch will slow evaporation of moisture while discouraging weed growth. Adding 2 - 4 inches of organic material such as compost or bark mulch will increase the ability of the soil to retain moisture. Press the mulch down around the dripline of each plant to form a slight depression which will prevent or minimize water runoff.

18. Don't water the gutter: Position your sprinklers so water lands on the lawn or garden, not on paved areas. Also, avoid watering on windy days.

19. Water your lawn only when it needs it: A good way to see if your lawn needs watering is to step on the grass. If it springs back up when you move, it doesn't need water. If it stays flat, the lawn is ready for watering. Letting the grass grow taller (to 3") will also promote water retention in the soil.

    Most lawns only need about 1" of water each week. During dry spells, you can stop watering altogether and the lawn will go brown and dormant. Once cooler weather arrives, the morning dew and rainfall will bring the lawn back to its usual vigor. This may result in a brown summer lawn, but it saves a lot of water.

20. Deep-soak your lawn: When watering the lawn, do it long enough for the moisture to soak down to the roots where it will do the most good. A light sprinkling can evaporate quickly and tends to encourage shallow root systems. Put an empty tuna can on your lawn - when it's full, you've watered about the right amount. Visit our natural lawn care page for more information.
21. Water during the early parts of the day; avoid watering when it's windy: Early morning is generally better than dusk since it helps prevent the growth of fungus. Early watering, and late watering, also reduce water loss to evaporation. Watering early in the day is also the best defence against slugs and other garden pests. Try not to water when it's windy - wind can blow sprinklers off target and speed evaporation.
22. Add organic matter and use efficient watering systems for shrubs, flower beds and lawns: Adding organic material to your soil will help increase its absorption and water retention. Areas which are already planted can be 'top dressed' with compost or organic matter.

You can greatly reduce the amount of water used for shrubs, beds and lawns by:

- the strategic placement of soaker hoses
- installing a rain barrel water catchment system
- installing a simple drip-irrigation system

Avoid over-watering plants and shrubs, as this can actually diminish plant health and cause yellowing of the leaves.

When hand watering, use a variable spray nozzle for targeted watering.

23. Don't run the hose while washing your car: Clean the car using a pail of soapy water. Use the hose only for rinsing - this simple practice can save as much as 150 gallons when washing a car. Use a spray nozzle when rinsing for more efficient use of water. Better yet, use a waterless car washing system; there are several brands, such as EcoTouch, which are now on the market.
24. Use a broom, not a hose, to clean driveways and sidewalks.
25. Check for leaks in pipes, hoses, faucets and couplings: Leaks outside the house may not seem as bad since they're not as visible. But they can be just as wasteful as leaks indoors. Check frequently to keep them drip-free. Use hose washers at spigots and hose connections to eliminate leaks.

Water conservation comes naturally when everyone in the family is aware of its importance, and parents take the time to teach children some of the simple water-saving methods around the home which can make a big difference.

Water Conservation Summary

In 1990, 30 states in the US reported 'water-stress' conditions. In 2000, the number of states reporting water-stress rose to 40. In 2009, the number rose to 45. There is a worsening trend in water supply nationwide. Taking measures at home to conserve water not only saves you money, it also is of benefit to the greater community.

Saving water at home does not require any significant cost outlay. Although there are water-saving appliances and water conservation systems such as rain barrels, drip irrigation and on-demand water heaters which are more expensive, the bulk of water saving methods can be achieved at little cost. For example, 75% of water used indoors is in the bathroom, and 25% of this is for the toilet. The average toilet uses 4 gallons per flush (gpf). You can invest in a ULF (ultra-low flush) toilet which will use only 2 gpf. But you can also install a simple tank bank, costing about $2, which will save .8 gpf. This saves 40% of what you would save with the ULF toilet. Using simple methods like tank banks, low-flow showerheads and faucet aerators you can retrofit your home for under$50.

By using water-saving features you can reduce your in-home water use by 35%. This means the average household, which uses 130,000 gallons per year, coulod save 44,00 gallons of water per year. On a daily basis, the average household, using 350 gallons per day, could save 125 gallons of water per day. The average individual, currently using 70 gallons per day, could save 25 gallons of water per day. When buying low-flow aerators, be sure to read the label for the actual 'gpm' (gallons per minute) rating. Often, the big box retailers promote "low-flow" which are rated at 2.5 gpm, which is at the top of the low-flow spectrum. This may be needed for the kitchen sink, but we find that a 1.5 gpm aerator works fine for the bathroom sink and most water outlets, delivering the same spray force in a comfortable, soft stream. Finally, it should be noted that installing low-flow aerators, showerheads, tank banks and other water-saving devices usually is a very simple operation which can be done by the homeowner and does not even require the use of tools. Water conservation at home is one of the easiest measures to put in place, and saving water should become part of everday family practice.

Water Conservation Solutions / Water is Life

Because of these circumstances, resources of consumable water are very small. The closest estimate to date is that there is about 1 per cent of water on the earth that is suitable for drinking. This is serious, because in many ways water represents life. Albert Szent-Gyorgyi, the Hungarian biochemist and Nobel Prize winner for medicine says, "Water is life's matter and matrix, mother and medium. There is no life without water." Consumable and fresh water is the need of every living being on earth. Humans, animals, other creatures, and agriculture all need water to survive and thrive. Without water our planet would become a vacant and barren wasteland.

And if you are thinking that water scarcity is just an issue in third world countries…think again. "Twenty-one percent of irrigation in the United States is achieved by pumping groundwater at rates that exceed the water's ability to recharge," wrote water expert Tony Clarke of the Polaris Institute.

Having water available in its clean and purest form is a blessing, and should be treated with the utmost respect. It is our obligation to save and conserve water as much as we can. In this way we will be helping ourselves and all of humanity.

Below are simple water conservation solutions. And while you practice these water conservation methods give a silent expression of thanks for having the clean water that so many others do not.

Water Conservation Solutions / Simple Things You Can Do

- Take shorter showers.
- Close the tap at once after using the water.
- While brushing, only open the tap when you need water.
- Do not over-water plants.
- Wash your car less frequently and use smaller amounts of water, or use no water when you wash your car.
- Fix leaky pipes immediately upon discovery.
- Slow the flow of water to consume less water, using less energy.
- Operate the washing machine only when you have a full load of laundry.
- Don't throw towels in the laundry after only one use.
- Run your dishwasher only when you have a full load.

These are some ideas to conserve water. Now, let's look at what solutions are available on the market that could help you use less water.

Water Conservation Solutions / Products Available that Help the Cause

- Use the water efficient taps. These taps turn off automatically when not in use.
- Buy appliances that are marked with water efficiency labels. These energy efficient machines use about half the amount of the water as standard appliances.
- Use water efficient showerheads.
- Install a dual flush toilet.
- Install aerators, flow regulators, flow restrictors, and pressure-limiting valves to your taps. These measures will minimize the flow of water considerably.
- Purchase washing machines with front-load washing. These machines use much smaller amounts of water compared to the other machines.
- Keep your appliances at the recommended level to ensure greater efficiency.
- Don't use water to wash your car. Yes, you read that correctly. Find out about this innovative new product at Freedom Waterless and see how it works.

Consider each of these points as a thank you to the planet. When individuals begin to take part in creating a healthier environment, the law of cause and effect will start to spread in a life-enhancing way.

3

METHODS AND TECHNIQUES FOR GROUND WATER RECHARGE

In rural areas, rain water harvesting is taken up considering watershed as a unit. Surface spreading techniques are common since space for such systems is available in plenty and quantity of recharged water is also large. Following techniques may be adopted to save water going waste through slopes, rivers, rivulets and nalas.

GULLY PLUG

Gully plugs are built using local stones, clay and bushes across small gullies and streams running down the hill slopes carrying drainage to tiny catchments during rainy season.

Gully Plugs help in conservation of soil and moisture.

The sites for gully plugs may be chosen whenever there is a local break in slope to permit accumulation of adequate water behind the bunds.

Contour bund

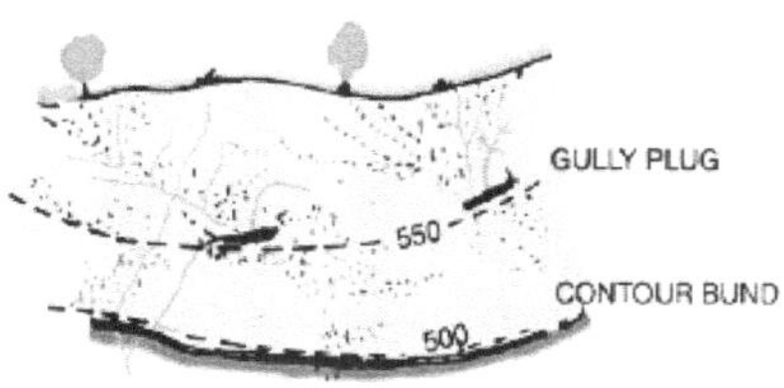

Contour bunds are effective methods to conserve soil moisture in watershed for long duration.

These are suitable in low rain fall areas where monsoon run off can be impounded by constructing bunds on the sloping ground all along the contour of equal elevation.

Flowing water is intercepted before it attains the erosive velocity by keeping suitable spacing between bunds.

Spacing between two contour bunds depends on the slope, the area and the permeability of the soil. Lesser the permeability of soil, the close should be spacing of bunds.

Contour bunding is suitable on lands with moderate slopes without involving terracing.

Gabion structure

This is a kind of check dam commonly constructed across small streams to conserve stream flows with practically no submergence beyond stream course.

A small bund across the stream is made by putting locally available boulders in a mesh of steel wires and anchored to the stream banks.

The height of such structures is around 0.5 m and is normally used in the streams with width of less than 10 m.

The excess water over flows this structure storing some water to serve as source of recharge. The silt content of stream water in due course is deposited in the interstices of the boulders. With the growth of vegetation, the bund becomes quite impermeable and helps in retaining surface water run off for sufficient time after rains to recharge the ground water body.

Gabion structure1 Gabion structure2

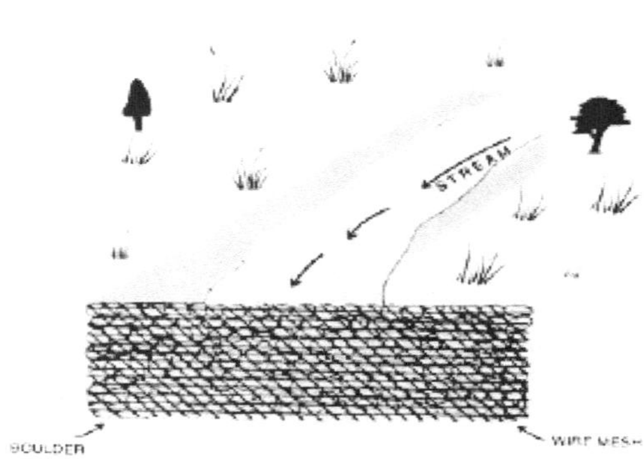

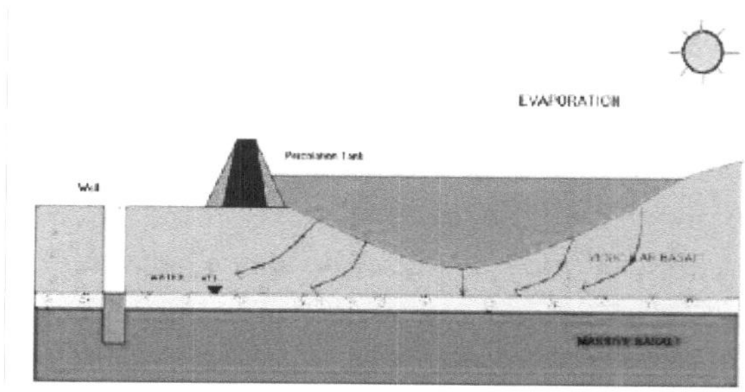

Percolation tank is an artificially created surface water body, submerging in its reservoir a highly permeable land, so that surface runoff is made to percolate and recharge the ground water storage.

Percolation tank should be constructed preferably on second to third order steams, located on highly fractured and weathered rocks, which have lateral continuity down stream.

The recharge area down stream should have sufficient number of wells and cultivable land to benefit from the augmented ground water.

The size of percolation tank should be governed by percolation capacity of strata in the tank bed. Normally percolation tanks are designed for storage capacity of 0.1 to 0.5 MCM. It is necessary to design the tank to provide a ponded water column generally between 3 & 4.5 m.

The percolation tanks are mostly earthen dams with masonry structure only for spillway. The purpose of the percolation tanks is to recharge the ground water storage and hence seepage below the seat of the bed is permissible. For dams upto 4.5 m height, cut off trenches are not necessary and keying and benching between the dam seat and the natural ground is sufficient.

Check dams / cement plugs / nala bunds

Check dams are constructed across small streams having gentle slope. The site selected should have sufficient thickness of permeable bed or weathered formation to facilitate recharge of stored water within short span of time.

The water stored in these structures is mostly confined to stream course and the height is normally less than 2 m and excess water is allowed to flow over the wall. In order to avoid scouring from excess run off, water cushions are provided at downstream side.

To harness the maximum run off in the stream, series of such check dams can be constructed to have recharge on regional scale.

Clay filled cement bags arranged as a wall are also being successfully used as a barrier across small nalas. At places, shallow trench is excavated across the nala and asbestos sheets are put on two sides. The space between the rows of asbestos sheets across the nala is backfilled with clay. Thus a low cost check dam is created. On the upstream side clay filled cement bags can be stacked in a slope to provide stability to the structure.

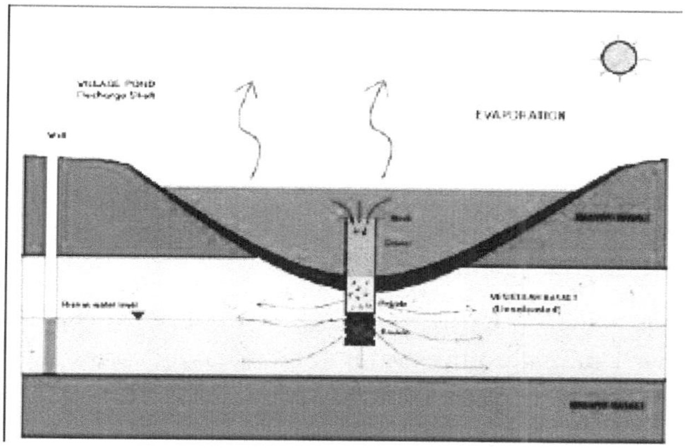

RECHARGE SHAFT

This is the most efficient and cost effective technique to recharge unconfined aquifer overlain by poorly permeable strata.

Recharge shaft may be dug manually if the strata is of non-caving nature. The diameter of shaft is normally more than 2 m.

The shaft should end in more permeable strata below the top impermeable strata. It may not touch water table.

The unlined shaft should be backfilled, initially with boulders/ cobbles followed by gravel and coarse sand.

In case of lined shaft the recharge water may be fed through a smaller conductor pipe reaching up to the filter pack.

These recharge structures are very useful for village ponds where shallow clay layer impedes the infiltration of water to the aquifer.

It is seen that in rainy season village tanks are fully filled up but water from these tanks does not percolate down due to siltation and tubewell and dugwells located nearby remains dried up. The water from village tanks get evaporated and is not available for the beneficial use.

By constructing recharge shaft in tanks, surplus water can be recharged to ground water. Recharge shafts of 0.5 to 3 m. diameter and 10 to 15 m. deep are constructed depending upon availability of quantum of water. The top of shaft is kept above the tank bed level preferably at half of full supply level. These are back filled with boulders, gravels and coarse sand.

In upper portion of 1 or 2 m depth, the brick masonry work is carried out for the stability of the structure.

Through this technique all the accumulated water in village tank above 50% full supply level would be recharged to ground water. Sufficient water will continue to remain in tank for domestic use after recharge.

DUGWELL RECHARGE

Existing and abandoned dug wells may be utilized as recharge structure after cleaning and desilting the same.

The recharge water is guided through a pipe from desilting chamber to the bottom of well or below the water level to avoid scouring of bottom and entrapment of air bubbles in the aquifer.

Recharge water should be silt free and for removing the silt contents, the runoff water should pass either through a desilting chamber or filter chamber.

Periodic chlorination should be done for controlling the bacteriological contaminations.

GROUND WATER DAMS OR SUB-SURFACE DYKES

Sub surface dyke or under-ground dam is a subsurface barrier across stream which retards the base flow and stores water upstream below ground surface. By doing so, the water levels in upstream part of ground water dam rises saturating otherwise dry part of aquifer.

The site where sub-surface dyke is proposed should have shallow impervious layer with wide valley and narrow out let.

After selection of suitable site, a trench of 1-2 m wide is dug across the breadth of stream down to impermeable bed. The trench may be filled with clay or brick/ concrete wall upto 0.5m. below the ground level.

For ensuring total imperviousness, PVC sheets of 3000 PSI tearing strength at 400 to 600 gauge or low-density polythene film of 200 gauges can also be used to cover the cut out dyke faces.

Since the water is stored within the aquifer, submergence of land can be avoided and land above the reservoir can be utilized even after the construction of the dam. No evaporation loss from the reservoir and no siltation in the reservoir takes place. The potential disaster like collapse of the dams can also be avoided.

GROUND WATER RECHARGE IN URBAN AREAS

In rural areas, rain water harvesting is taken up considering watershed as a unit. Surface spreading techniques are common since space for such systems is available in plenty and quantity of recharged water is also large. Following techniques may be adopted to save water going waste through slopes, rivers, rivulets and nalas.

In urban areas, rain water available from roof tops of buildings, paved and unpaved areas goes waste. This water can be recharged to aquifer and can be utilized gainfully at the time of need. The rain water harvesting system needs to be designed in a way that it does not occupy large space for collection and recharge system. A few techniques of roof top rain water harvesting in urban areas are described below.

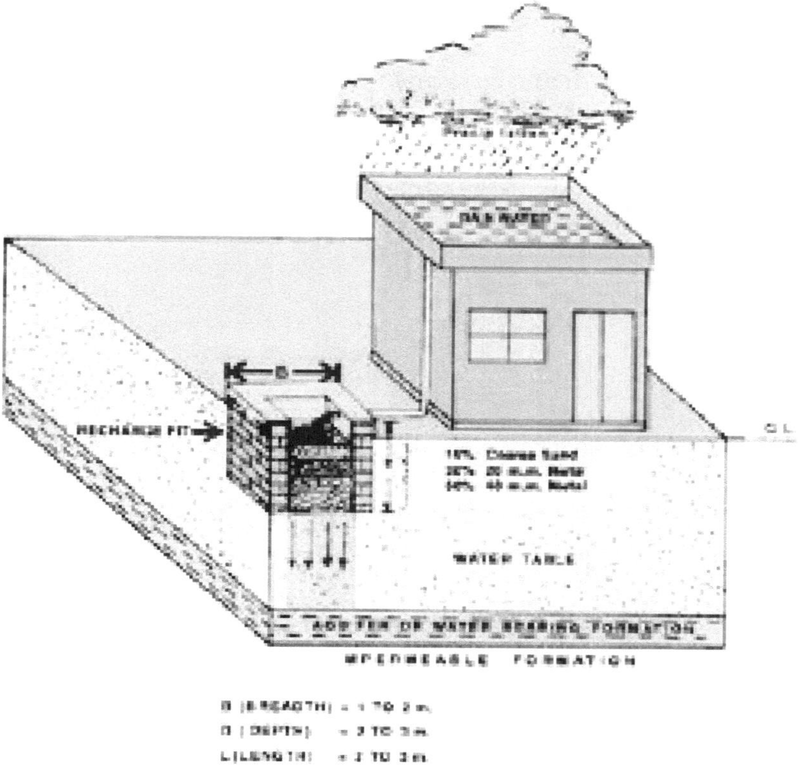

REACHARGE PIT

In alluvial areas where permeable rocks are exposed on the land surface or are located at very shallow depth, rain water harvesting can be done through recharge pits.

The technique is suitable for buildings having a roof area of 100 sq.m. These are constructed for recharging the shallow aquifers.

Recharge Pits may be of any shape and size. They are generally constructed 1 to 2 m. wide and 2 to 3 m deep. The pits are filled with boulders (5-20 cm), gravels (5-10mm) and coarse sand (1.5- 2mm) in graded form. Boulders at the bottom, gravels in between and coarse sand at the top so that the silt content that will come with runoff water will be deposited on the top of the coarse sand layer and can easily be removed. For smaller roof area, pit may be filled with broken bricks/ cobbles.

A mesh should be provided at the roof so that leaves or any other solid waste / debris is prevented from entering the pit. A desilting /collection chamber may also be provided at the ground to arrest the flow of finer particles to the recharge pit.

The top layer of sand should be cleaned periodically to maintain the recharge rate.

By-pass arrangement is to be provided before the collection chamber to reject the first showers.

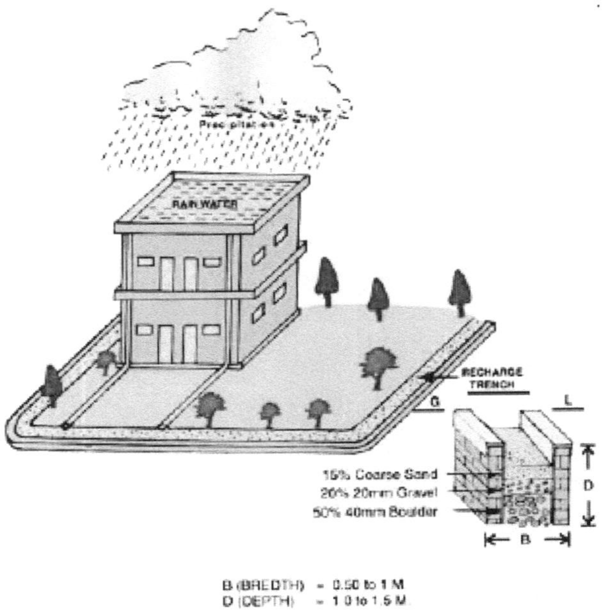

Recharge trench

Recharge trenches are suitable for buildings having roof area of 200-300 sq. m. and where permeable strata is available at shallow depths.

Trench may be 0.5 to 1 m wide, 1 to 1.5m. deep and 10 to 20 m. long depending upon availability of water to be recharge.

These are back filled with boulders (5-20cm), gravel (5-10 mm) and coarse sand (1.5-2 mm) in graded form – boulders at the bottom, gravel in between and coarse sand at the top so that the silt content that will come with runoff will be coarse sand at the top of the sand layer and can easily be removed.

A mesh should be provided at the roof so that leaves or any other solid waste/debris is prevented from entering the trenches and a desilting/collection chamber may also be provided on ground to arrest the flow of finer particles to the trench.

By-pass arrangement is to be provided before the collection chamber to reject the first showers.

The top layer of sand should be cleaned periodically to maintain the recharge rate.

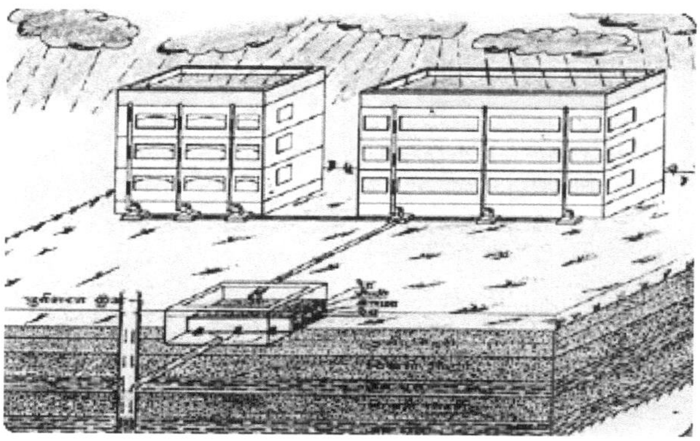

TUBE WELLS

In areas where the shallow aquifers have dried up and existing tubewells are tapping deeper aquifer, rain water harvesting through existing tubewell can be adopted to recharge the deeper aquifers.

PVC pipes of 10 cm dia are connected to roof drains to collect rainwater. The first roof runoff is let off through the bottom of drainpipe. After closing the bottom pipe, the rainwater of subsequent rain showers is taken through a T to an online PVC filter. The filter may be provided before water enters the tubewells. The filter is 1 –1.2 m. in length and is made up of PVC pipe. It's diameter should vary depending on the area of roof, 15 cm if roof area is less than 150 sq m and 20 cm if the roof area is more. The filter is provided with a reducer of 6.25 cm on both the sides. Filter is divided into three chambers by PVC screens so that filter material is not mixed up. The first chamber is filled up with gravel (6-10mm), middle chamber with pebbles (12-20 mm) and last chamber with bigger pebbles (20-40 mm).

If the roof area is more, a filter pit may be provided. Rainwater from roofs is taken to collection/desilting chambers located on ground. These collection chambers are interconnected as well as connected to the filter pit through pipes having a slope of 1:15. The filter pit may vary in shape and size depending upon available runoff and are back-filled with graded material, boulder at the bottom, gravel in the middle and sand at the top with varying thickness (0.30-0.50m) and may be separated by screen. The pit is divided into two chambers, filter material in one chamber and other chamber is kept empty to accommodate excess filtered water and to monitor the quality of filtered water. A connecting pipe with recharge well is provided at the bottom of the pit for recharging of filtered water through well.

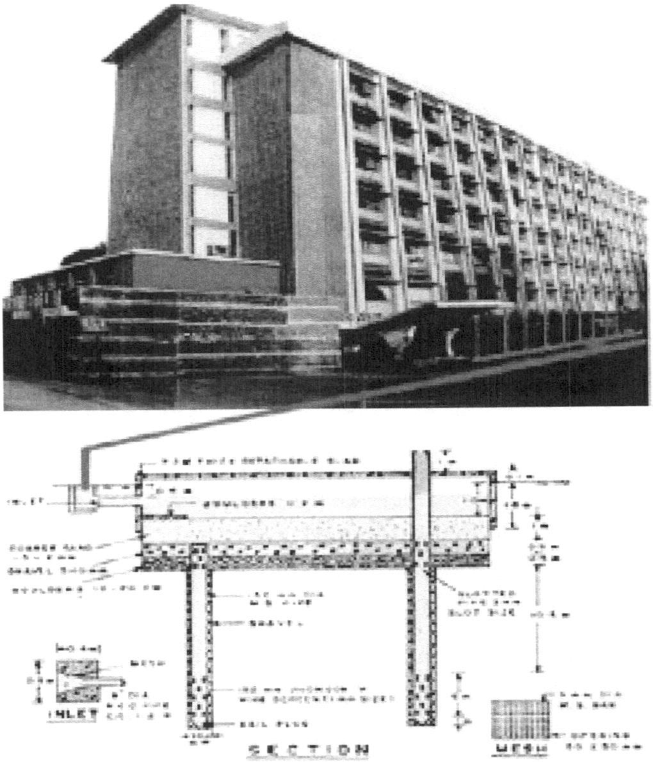

Trench with recharge well

In areas where the surface soil is impervious and large quantities of roof water or surface runoff is available within a very short period of heavy rainfall, the use of trench/ pits is made to store the water in a filter media and subsequently recharge to ground water through specially constructed recharge wells.

This techniques is ideally suited for area where permeable horizon is within 3m below ground level.

Recharge well of 100-300 diameter is constructed to a depth of at least 3 to 5 m below the water level. Based on the lithology of the area, well assembly is designed with slotted pipe against the shallow and deeper aquifer.

A lateral trench of 1.5 to 3m width and 10 to 30 m length, depending upon the availability of water is constructed with the recharge well in the centre.

The number of recharge wells in the trench can be decided on the basis of water availability and local vertical permeability of the rocks.

The trench is backfilled with boulders, gravels and coarse sand to act as a filter media for the recharge wells.

If the aquifer is available at greater depth say more than 20 m, a shallow shaft of 2 to 5 m diameter and 3-5 metres deep may be constructed depending upon availability of runoff. Inside the shaft a recharge well of 100-300 mm dia is constructed for recharging the available water to the deeper aquifers. At the bottom of the shaft a filter media is provided to avoid choking of recharge well.

Water strategy: Ponds need pondering

In the last few years, the water situation in Maharashtra has got worse resulting in severe droughts leading to drinking water scarcity and agricultural crisis. This has caused immense suffering for the rural folk in the state and saw instances of violence in the name of water. The government was forced to enforce Section 144 of the Indian Penal Code to facilitate smooth distribution of water among the population.

This article, Problematic uses and practices of farm ponds in Maharashtra published in the Economic and Political Weekly says that the situation was so bad that it was for the first time in the history of the state that water trains were used to supply potable water in drought-hit villages.

What has led to this situation in the state?

The article says that this grave water situation has to do with the widespread electrification in the farms following the green revolution. This resulted in deep borewell technologies which led to increased extraction of groundwater from wells and surface water storages. As borewell technology became cheaper and easily available, small marginal, as well as large farmers, got caught in the race to dig deeper in search of water, thus depleting the groundwater resources at alarming rates. Now the situation is such that even after digging deeper, farmers are unable to procure groundwater.

Farm ponds as a solution

Farm pond technology has been hailed as a ray of hope or a miracle strategy to deal with this increasing water crisis in the state. Farm ponds are square or rectangular holes made on the earth which harvest rainwater and store it for future use. The farm pond has an inlet that regulates the flow of water inside the pond while the outlet discharges excess water. The pond is surrounded by a small bund which prevents erosion from the banks of the pond. Water from the farm pond can be used for the fields either manually or by pumping or both.

Realising the advantages of the farm pond strategy to mitigate drought situations in Maharashtra, both the state and the central governments have come up with various schemes to promote and subsidise the construction of the ponds on a large scale. The central government budget for 2016–17 has set an ambitious target to construct five lakh farm ponds and wells within a year in rain water-scarce areas of the country.

PROBLEMATIC FARM POND PRACTICES

The article, however, argues that there are major problems in the actual implementation of the farm pond strategy at the ground level. There are gaps between the proposed objectives of the farm ponds and its actual use by farmers.

For example, although the important function of a farm pond is to harvest rainwater, most farm pond owners extract groundwater from dug wells and borewells and then store it in farm ponds, thus defeating the very purpose of its construction.

Moreover, there are major faults in the way the farm ponds are constructed like the absence of inlet and outlet valves to receive and discharge rainwater. Farm ponds help groundwater recharge through percolation of the stored rainwater. However, in practice, high micron plastic paper is applied in almost all functional farm ponds to stop the seepage of stored water. This reduces the possibility of the percolation of water from the farm pond to recharge groundwater.

There seems to be a lack of clarity at the policy level on the purpose and strategies used for constructing farm ponds. There is inadequate planning while sanctioning the construction of farm ponds and poor regulatory mechanisms in their use and practice. No attempts are being made to understand the overall water resources available at the village level or watershed area under consideration, the total carrying capacity of that area which supplies water for different uses and structures while planning and deciding on the number, the size and the depth of farm ponds. Evaporation of water from the farm ponds is another serious issue that is not considered.

The article thus argues that although the construction of large-sized farm ponds has helped to raise the income of some farmers who are financially better off, it has worsened the overall groundwater situation in water-scarce areas due to the unregulated extraction of groundwater. In a few villages, people still face severe drinking water problem and depend on public tankers for drinking water security while farm ponds in the nearby fields are full of extracted groundwater.

Besides depletion of groundwater, this has also led to drying up of dug wells and borewells in the neighbouring area plus a reduction in runoff leading to water scarcity in downstream villages. Pumping large amounts of groundwater from shallow and deep aquifers for storing in farm ponds may also affect the water ?ows in streams and drains by threatening the already strained ecosystem.

What can be the way out?

The article ends by making the following policy recommendations to effectively deal with the situation:

The need for a ban on extraction of groundwater for storage in farm ponds in notified areas, exploited watersheds, and water scarcity zones.

Setting limits on the number of farm ponds that can be constructed in a village or watershed area taking into consideration the carrying capacity of the village/watershed.

Controlling the size of farm ponds and ensuring that farmers abide by the sanctioned design appropriate for the village or the watershed area.

Focus on small farmers for the provision of subsidy.

Explore alternatives to the plastic lining currently used for lining the farm ponds.

Encourage construction of common farm ponds and community sharing to secure drinking water needs and better water management.

Explore strategies to reduce the rate of evaporation of water stored in the farm ponds.

COMPONENTS OF RAINWATER HARVESTING SYSTEM

The common components of a rainwater harvesting system are:

1. Catchments

The surface which directly receives the rainfall and provides water to the system is called catchment area. It can be a paved area like a terrace or courtyard of a building, or an unpaved area like a lawn or open ground. A roof made of reinforced cement concrete (RCC), galvanized iron or corrugated sheets can also be used for water harvesting.Components of Rainwater Harvesting System

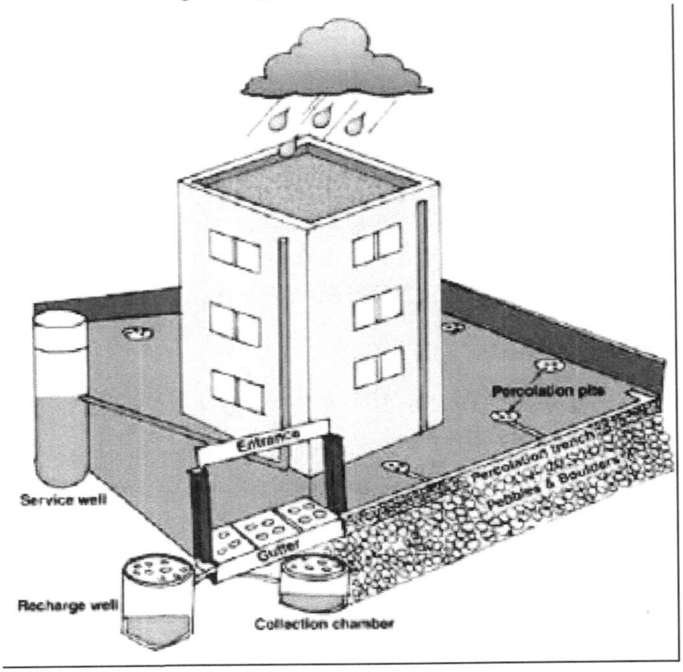

2. Coarse Mesh

It prevents the passage of debris, provided in the roof.

3. Gutters

Channels which surrounds edge of a sloping roof to collect and transport rainwater to the storage tank. Gutters can be semi - circular or rectangular and mostly made locally from plain galvanized iron sheet. Gutters need to be supported so they do not sag or fall off when loaded with water. The way in which gutters are fixed mainly depends on the construction of the house, mostly iron or timber brackets are fixed into the walls.

4. Conduits

Conduits are pipelines or drains that carry rainwater from the catchment or roof top area to the harvesting system. Commonly available conduits are made up of material like polyvinyl chloride (PVC) or galvanized iron (GI).

5. First-flushing

A first flush device is a valve which ensures flushing out of first spell of rain away from the storage tank that carries a relatively larger amount of pollutants from the air and catchment surface.

6. Filters

The filter is used to remove suspended pollutants from rainwater collected from roof top water. The Various types of filters generally used for commercial purpose are Charcoal water filter, Sand filters, Horizontal roughing filter and slow sand filter.

7. Storage facility

There are various options available for the construction of these tanks with respect to the shape, size, material of construction and the position of tank and they are: -Shape : Cylindrical, square and rectangular. Material of construction: Reinforced cement concrete (RCC), masonry, Ferrocement etc. Position of tank: Depending on land space availability these tanks could be constructed above ground, partly underground or fully underground. Some maintenance measures like disinfection and cleaning are required to ensure the quality of water stored in the container. If harvested water is decided to recharge the underground aquifer/ reservoir, then some of the structures mentioned below are used.

8. Recharge structures

Rainwater Harvested can also be used for charging the groundwater aquifers through suitable structures like dugwells, borewells, recharge trenches and recharge pits. Various recharge structures are possible - some which promote the percolation of water through soil strata at shallower depth (e.g., recharge trenches, permeable pavements) whereas others conduct water to greater depths from where it joins the groundwater (e.g. recharge wells). At many locations, existing structures like wells, pits and tanks can be modified as recharge structures, eliminating the need to construct any fresh structures. Some of the few commonly used recharging methods are recharging of dug wells and abandoned tube wells, Settlement tank, Recharging of service tube wells, Recharge pits, Soakaways / Percolation pit , Recharge troughs, Recharge trenches, Modified injection well.

DESIGN TIPS FOR COLLECTION OF RAINWATER

Water collected from roof can be a) stored for direct use b) allowed to infiltrate as groundwater recharge and c) combination of part storage and part recharge. There are three structural components a) roof water collection system b) storage tank and c) recharge pit. The cost of the system will depends upon the cost of each of these components. Water collected from the early showers are allowed to flow away since they contain some dirt. Once the roof top is washed clean by rainfall, the remaining water is collected. Nearly 80% of the rain fall can be easily harvested. It is much easy to collect water from a sloping roof top simply by placing lateral gutters (semi-circular collecting pipe) to bring the collected water at one common point. Flat, concrete roof tops sometimes offer more serious challenges for collection of rain water. Usually flat concrete roof tops are provided with at least 4 outlets (drain pipes) running along the four corners. This means the water collected from all the 4 corners pipes are to be brought at the common storage site through a network of additional pipe lines. This may spoil the aesthetics and increase cost. Also, there is a need to keep the roof top clean to prevent pollution of rain water by preventing people to use the roof top for various purposes including storing of junk items.

DESIGN TIPS FOR STORAGE TANK

Storage tanks could be constructed a) underground b) half underground and half over ground and c) over-ground depending upon its size and availability of land area. Storage tanks could be made of a) PVC (ready made Syntex type), b) Ferro-cement tanks and c) cement concrete tanks. Cost not being a consideration, a) for small storage (<> 20,000 L), cement concrete tanks would be ideal. Ferro-cement tanks are of least cost. These are tanks made from pasting rich cement mortar over a tank structure made from bending, shaping and binding 5 mm or so iron rods as reinforcement with an inside and out side layer of chicken mess (common wire net). Unlike cement concrete tanks, these are less prone to cracks/leakages and can be made in any shape (round, cylindrical etc.). Masons having experience in working with ferro cement are required to undertake this job. Many NGOs have such masons. Storage tank to be used for drinking water must have a filter system at the entry point, a water extraction system for taking out water (tap, hand pump) and a draining system for cleaning the tank (manhole, drain pipe etc.) periodically. A diversion arrangement (T-joint with valve) is required at the outlet of roof collected water pipe line to allow the water to be either disposed off (dirty water after initial rain) or stored in the storage tank through the filter system for use as drinking water. Rain water is generally deficient in dissolved minerals necessary for human body. Some scientists have suggested that on prolonged use, low Calcium (Ca) in rain water may absorb some Ca from our body making our bones weak (also applies to some RO purified water). However, so far there has been no report of any adverse effect of using rain water as drinking water.

Design Tips for Groundwater Recharge

Artificial groundwater recharge is professed when a) there is excess source water available at site and b) quality of source water is compatible with the groundwater to be recharged. It is always safe and desirable to construct a RWH structure, where water is partly stored for direct use and partly allowed to percolate to recharge groundwater. This helps in maintaining quality control of the water used for recharge. The recharge structure is usually a 1- 2 cubic meter pit (could be larger when large quantity of water is to be recharged), filled with graded filter materials like cobbles, pebbles, gravel and sand in layers. The Objective is to let silt free good quality water to percolate. Designing site specific filter beds are of critical importance. If the interstices of the filter are too small, water percolation will be slow. Conversely, filters with larger pores will allow faster percolation but water may not be free from suspended matters. For this reason, it is customary to have a pre-filter storage tank from where water is supplied to the pit as per its percolation capacity. In hard rock areas (basalt and granite) where water occurs under water table condition in the weathered formation, there is no need to attach an extension pipe from the bottom of the recharge pit. Bur in areas where water occurs under water table condition at a deeper level (20 – 100 m bgl), it would be necessary to attach an extension pipe. The purpose of the pipe is to allow the source water to join directly to the aquifer (water table) without being lost in the over-lying unsaturated zone. Ideally, the length of the pipe should be such that it ends near the water table which however is known to fluctuate seasonally. It is necessary, that the bottom most part of the pipe be a slotted pipe. Similarly, if the pipe extends upwards to the ground level, the top most pipe should also be a slotted pipe. An abandoned (dry but clean) dug well in hard rock area can be used as an effective recharge pit provided, a) water table is not too far below from the bottom of the well and b) formation near the well bottom is porous and permeable and c) the water diverted in to the well for recharge is of good quality. Such recharge practice should be best done in consultation with a geohydrologist and better be avoided if there is a risk of contaminated water entering in to the well and finally to the groundwater. Artificial groundwater recharge through a number of such recharge structures will not only increase water availability (raised water table) but will also improve water quality. If properly constructed and managed, it is possible to use a bore well both as a recharge well during the monsoon period and supply (abstraction) well during summer months. All RWH structure will require regular monitoring and periodical (at least annual) cleaning and maintenance of its various components. It is advisable to rather not construct these RWH structures, if regular maintenance cannot be assured.

4

RAIN WATER HARVESTING PRACTICES

Water is a common substance that is essential to all known forms of life. Water covers 70 percent of the planet's surface. On the planet Earth water is mostly found in oceans and large bodies of water. 1.6 percent of the Earth's water is found underground in aquifers. A tiny percentage of the world's water is found in the bodies of plants and animals and in manufactured goods.

Humans and other animals need access to a supply of fresh and clean drinking water for their survival. The availability of safe drinking water in almost all areas of the world has increased to a large extent over the last few decades. Despite this fact some observers have interpolated that by 2025 a large part of the world's population will be facing shortages of water. One household in the USA for example uses on average 127,000 gallons of water per year. But if greater use was made of low-flow fixtures on their taps, washing machine, shower head and toilet, one USA household could save 40,000 gallons of water per year and 30% on their water bill. If every household used low-flow fixtures, we could save 5.4 billion gallons/day which is worth $11.3 million dollars! The United Nations calls this situation a water crisis. The role of water in the world's economy is enormous because it is used as a dilutant for a wide variety of chemical substances in industrial and medical applications.

Agriculture is the biggest consumer of freshwater using over 70 percent of freshwater supplies. Recycling water is the use of treated and processed wastewater for useful purposes like agricultural irrigation, industrial processes, toilet flushing, and groundwater replenishment which is also called a groundwater recharge. Water recycling is also known as water reusing or water reclamation. These terms are often used interchangeably.

How Can Water Recycling Benefit Us?

Water recycling provides enormous environmental benefits. It also provides an additional source of water for various purposes. This a list of some benefits that water recycling can present.

- Water recycling decreases the extraction of water form sources that may be dwindling and may stop being viable as habitats for valuable and endangered wildlife.
- Recycling wastewater can decrease the discharge of effluents that may damage and pollute the ecosystems of the sensitive bodies of water.
- Recycled water can be used to create new wetlands or to enhance and improve the quality of existing ones.
- Water recycling can reduce and prevent pollution by leaving damaging pollutants at the treatment plant.

Recycling Water / What it Does For the Future

Recycling water has been established as an effective and safe way to provide non-potable water for a wide range of industrial, home, and agricultural use. The non-potable utilization of recycled water can only continue to grow. In numerous parts of the United States the uses of recycled water are increasing to accommodate the environmental situation and growing demands for water.

Technological advances in wastewater treatment and health studies in the use of indirect potable water have indicated the possible planned indirect potable reuse might soon become feasible. As the need for more water grows and environmental pressures increase, the necessity for water recycling can only increase as well. Water recycling will need to have a greater role in providing water for all the needs of society. We must work together to make water recycling and water conservation effective enough to help us sustain our valuable and vital water resources.

Water Conservation Facts and Tips

- Less than 2% of the Earth's water supply is fresh water.
- Of all the earth's water, 97% is salt water found in oceans and seas.
- Only 1% of the earth's water is available for drinking water. Two percent is frozen.

- The human body is about 75% water.
- A person can survive about a month without food, but only 5 to 7 days without water.
- Every day in the United States, we drink about 110 million gallons of water.
- Landscaping accounts for about half the water Californians use at home. Showers account for another 18 percent, while toilets use about 20 percent.
- Showering and bathing are the largest indoor uses (27%) of water domestically.
- The average American uses 140-170 gallons of water per day.
- If every household in America had a faucet that dripped once each second, 928 million gallons of water a day would leak away.
- There are 7.48 gallons in a cubic foot of water. Therefore, 2000 cubic feet of water is 14,960 gallons.
- An acre foot of water is about 326,000 gallons. One-half acre foot is enough to meet the needs of a typical family for a year. There are 7.48 gallons in a cubic foot of water.
- It takes 3.3 acre feet of water to grow enough food for an average family for a year.
- A leaky faucet can waste 100 gallons a day.
- One flush of the toilet uses 6 ½ gallons of water.
- An average bath requires 37 gallons of water.
- An average family of four uses 881 gallons of water per week just by flushing the toilet.
- The average 5-minute shower takes 15-25 gallons of water--around 40 gallons are used in 10 minutes.
- Take short showers instead of baths.
- A full bathtub requires about 36 gallons of water.
- You use about 5 gallons of water if you leave the water running while brushing your teeth.
- If you water your grass and trees more heavily, but less often, this saves water and builds stronger roots.

- Each person needs to drink about 2 ½quarts (80 ounces) of water every day.
- Water your lawn only when it needs it. If you step on the grass and it springs back up when you move, it doesn't need water. If it stays flat, it does need water.
- Run your dishwasher and washing machine only when they are full.
- When washing a car, use soap and water from a bucket. Use a hose with a shut-off nozzle for rinsing.
- Never put water down the drain when there may be another use for it such as watering a plant or garden, or cleaning.
- Avoid flushing the toilet unnecessarily. Dispose of tissues, insects and other such waste in the trash rather than the toilet.
- When washing dishes by hand, fill one sink or basin with soapy water. Quickly rinse under a slow-moving stream from the faucet.
- An automatic dishwasher uses 9 to 12 gallons of water while hand washing dishes can use up to 20 gallons.
- Store drinking water in the refrigerator rather than letting the tap run every time you want a cool glass of water.
- Water lawns during the early morning hours, or evening when temperatures and wind speed are the lowest.
- This reduces losses from evaporation.
- Do not hose down your driveway or sidewalk.
- Use a broom to clean leaves and other debris from these areas.
- Using a hose to clean a driveway wastes hundreds of gallons of water.
- Don't leave the water running when brushing your teeth or shaving. Get in the habit of turning off the water when it's not being used.
- Use of bowl of water to clean fruits & vegetables rather than running water over them. You can reuse this for your house plants.
- Public water suppliers process 38 billion gallons of water per day for domestic and public use.
- Approximately 1 million miles of pipelines and aqueducts carry water in the U.S. & Canada. That's enough pipe to circle the earth 40 times.
- About 800,000 water wells are drilled each year in the United States for domestic, farming, commercial, and water testing purposes.

- More than 13 million households get their water from their own private wells and are responsible for treating and pumping the water themselves.
- Industries released 197 million pounds of toxic chemicals into waterways in 1990.
- You can refill an 8-oz glass of water approximately 15,000 times for the same cost as a six-pack of soda pop.
- A dairy cow must drink four gallons of water to produce one gallon of milk.
- 300 million gallons of water are needed to produce a single day's supply of U.S. newsprint.
- One inch of rainfall drops 7,000 gallons or nearly 30 tons of water on a 60' by 180' piece of land.

FACTS ON SAVING WATER

According to the United States Environmental Protection Agency (EPA), a family of four uses 400 gallons of water every day. Despite the majority of the Earth being covered in water, only 1 percent of it is useable - the rest is salty ocean water or permanently frozen. Water is a limited resource, and with the population continuing to grow, it is essential to practice water conservation. Here are some water conservation facts and ways to reduce the amount of water you use or waste.

Water Conservation Facts

Many of your everyday activities use water - from bathing and cleaning the house to drinking and cooking. On a broader scale, water is necessary for agriculture, power plants and industrial or mining. The American lifestyle requires a great deal of this precious resource. Here are some surprising water usage statistics from the EPA.

- The average American family spends $500 per year on its water and sewer bill.
- Treating and delivering water takes a lot of energy - letting your faucet run for 5 minutes takes as much energy as letting a 60-watt lightbulb run for 14 hours.
- More than 30 percent of household water is for outdoor use, such as watering lawns.
- Turning off the tap water while you brush your teeth can save 8 gallons of water a day - that equals about 200 gallons of water saved per month.

- A faucet that leaks 1 drip per second wastes 3,000 gallons of water each year.
- Leaky toilets can waste up to 200 gallons per day, the equivalent of flushing the toilet 50 times for no reason.
- Older toilets use 60 percent more water per flush as compared to newer high-efficiency model (look for the WaterSense label when buying a new toilet).
- The average bath requires 70 gallons of water while a shower uses 10 to 25 gallons.
- A running hose outdoors - for washing cars or watering plants - can waste 6 gallons of water per minute when left unattended.
- US power plants use 136 billion gallons of water each day.
- US agriculture uses 142 billion gallons of water per day for irrigation.
- US industrial facilities require more than 20 billion gallons of water every day.
- Lower water levels in fresh water reservoirs can contribute to concentrated levels of both natural and human-based pollutants in the water supply.

WATER SAVING TIPS

Environmental advocate Lesley Chilcott, co-founder of Unscrew America and documentary filmmaker for former vice-president Al Gore's An Inconvenient Truth, recommends the following water-saving tips.

1. Shut the Water off

Instead of letting the water run while you brush your teeth, shave, peel vegetables or scrub dishes, turn it off until you need it.

2. Shower Efficiently

Don't run the shower for five minutes to warm it up while you are doing something else - adjust the temperature when you get in. And keep your shower time to five minutes. Turn the shower off if you need to shave or deep condition your hair.

3. Install Energy-efficient Appliances

Use an EnergyStar washing machine and dishwasher and only run full loads, or at the very least use the lowest water level setting you need for smaller loads.

4. Update your Toilet

Install a low-flow or high-efficiency toilet to save nearly 2 gallons of water per flush.

5. Change your Showerhead

In addition to shorter showers, install a low-flow showerhead, which will also reduce water waste. While you're at it, repair your leaky faucets.

6. Plant Sense

Change your landscaping and house plants to low-water, native or drought-tolerant plants. Then do your watering in the morning or evening when there is less chance of evaporation. Additionally, use your broom instead of the hose to clear debris from your driveway and side-walks.

7. Become a Flexitarian

The food production water usage for plant-based foods is significantly less than that of meat-based foods. A plant-based diet is healthier, too. (Click for more information on The Flexitarian Diet.)

8. Install a Water Filtration System

Instead of buying bottle after bottle of drinking water, install a water filter and use refillable beverage containers.

9. Use a Bucket

Wash your car with a bucket of soapy water instead of letting the hose run.

10. Set up a Rain Water Barrel

Collect rain water and use it later to water the garden. Another option is to install a gray water system.

11. Use Biodegradable Cleaning Products

Take advantage of the growing selection of eco-friendly household cleaning products, dish washing soaps, and laundry detergents. Biodegradable products won't pollute the water.

RAIN WATER HARVESTING

The term rainwater harvesting is being frequently used these days, however, the concept of water harvesting is not new for India. Water harvesting techniques had been evolved and developed centuries ago.

Ground water resource gets naturally recharged through percolation. But due to indiscriminate development and rapid urbainzation, exposed surface for soil has been reduced drastically with resultant reduction in percolation of rainwater, thereby depleting ground water resource. Rainwater harvesting is the process of augmenting the natural filtration of rainwater in to the underground formation by some artificial methods. "Conscious collection and storage of rainwater to cater to demands of water, for drinking, domestic purpose & irrigation is termed as Rainwater Harvesting."

Rainwater harvesting is the accumulation and storage of rainwater for reuse before it reaches the aquifer. Uses include water for garden, water for livestock, water for irrigation, cleaning of bathrooms as in first flush, etc. In many places the water collected is just redirected to a deep pit with percolation. The harvested water can be used for drinking water as well if the storage is a tank that can be accessed and cleaned when needed.

Rainwater harvesting provides an independent water supply during regional water restrictions, and in developed countries is often used to supplement the mains supply. Rainwater harvesting systems are appealing as they are easy to understand, install and operate. They are effective in 'green droughts' as water is captured from rainfall where runoff is insufficient to flow into dam storages. The quality of captured rainwater is usually sufficient for most household needs, reducing the need for detergents because rainwater is soft. Financial benefits to the users include that rain is 'renewable' at acceptable volumes despite climate change forecasts, and rainwater harvesting systems generally have low running costs, providing water at the point of consumption (Ferguson 2012).

Benefits of widespread rainwater harvesting to the regional reticulated supply system may include reduced treatment, pumping, operation and augmentation costs, reducing peak storm water runoff and storm water processing costs, as well as reduced greenhouse gas emissions due to reduced dependence on pumping and potential augmentation through sources such as desalination (Coombes 2007, White, 2009).

The principle of collecting and using precipitation from a catchments surface. An old technology is gaining popularity in a new way. Rain water harvesting is enjoying a renaissance of sorts in the world, but it traces its history to biblical times. Extensive rain water harvesting apparatus existed 4000 years ago in the Palestine and Greece. In ancient Rome, residences were built with individual cisterns and paved courtyards to capture rain water to augment water from city's aqueducts. As early as the third millennium BC, farming communities in Baluchistan and Kutch impounded rain water and used it for irrigation dams.

System of collection rainwater and conserving for future needs has traditionally been practiced in India. The traditional systems were time-tested wisdom of not only appropriate technology of Rainwater Harvesting, but also water management systems, where conservation of water was the prime concern. Traditional water harvesting systems were Bawaries, step wells, jhiries, lakes, tanks etc. These were the water storage bodies to domestic and irrigation demands. People were themselves responsible for maintenance to water sources and optimal use of water that could fulfill their needs.

ARTIFICAL RECHARGE TO GROUND WATER

Artificial recharge to ground water is a process by which the ground water reservoir is augmented at a rate exceeding that obtaining under natural conditions or replenishment. Any man-made scheme or facility that adds water to an aquifer may be considered to be an artificial recharge system.

Why Harvest Rainwater ?

Rain water harvesting is essential because :

Surface water is inadequate to meet our demand and we have to depend on ground water.

Due to rapid urbanization, infiltration of rain water into the sub-soil has decreased drastically and recharging of ground water has diminished.

As you read this guide, seriously consider conserving water by harvesting and managing this natural resource by artificially recharging the system. The examples covering several dozen installations successfully operating in India constructed and maintained by CGWB, provide an excellent snapshot of current systems.

This is perhaps one of the most frequently asked question, as to why one should harvest rainwater. There are many reasons but following are some of the important ones.

- To arrest ground water decline and augment ground water table
- To beneficiate water quality in aquifers
- To conserve surface water runoff during monsoon
- To reduce soil erosion
- To inculcate a culture of water conservation

How to harvest rainwater:

Broadly there are two ways of harvesting rainwater:

(i) Surface runoff harvesting

(ii) Roof top rainwater harvesting

SURFACE RUNOFF HARVESTING

In urban area rainwater flows away as surface runoff. This runoff could be caught and used for recharging aquifers by adopting appropriate methods.

Roof top rainwater harvesting (RTRWH)

It is a system of catching rainwater where it falls. In rooftop harvesting, the roof becomes the catchments, and the rainwater is collected from the roof of the house/building. It can either be stored in a tank or diverted to artificial recharge system. This method is less expensive and very effective and if implemented properly helps in augmenting the ground water level of the area.

Rainwater harvesting is a technology used for collecting and storing rainwater from rooftops, the land surface or rock catchments using simple techniques such as jars and pots as well as more complex techniques such as underground check dams. The techniques usually found in Asia and Africa arise from practices employed by ancient civilizations within these regions and still serve as a major source of drinking water supply in rural areas. Commonly used systems are constructed of three principal components; namely, the catchment area, the collection device, and the conveyance system.

Quality

The concentration of contaminants is reduced significantly by diverting the initial flow of runoff water to waste. Improved water quality can also be obtained by using a floating draw-off mechanism (rather than from the base of the tank) and by using a series of tanks, with draw from the last in series. The stored rainwater may need to be analyzed properly before use in a way appropriate to its safety.

System Setup

Rainwater harvesting systems can be installed with minimal skills. The system should be sized to meet the water demand throughout the dry season since it must be big enough to support daily water consumption. Specifically, the rainfall capturing area such as a building roof must be large enough to maintain adequate flow. Likewise, the water storage tank should be large enough to contain the captured water.

Vendors

There are three main types of companies operating in the rainwater harvesting industry: makers of water storage, makers of accessories, and integrators. Water storage companies make tanks, barrels, and underground cisterns. Accessories are added to facilitate or improve the water capturing process. Integrators are regional practitioners which install systems.

HISTORY

Early Era

Around the third century BCE, farming communities in Baluchistan (in present-day Pakistan, Afghanistan and Iran), and Kutch (in present-day India) used rainwater harvesting for irrigation. In ancient Tamil Nadu (India), rainwater harvesting were done by Chola kings. Rainwater from the Brihadeeswarar temple was collected in Sivaganga tank. During the later Chola period, the Vrnam tank was built (1011 to 1037 CE) in Cuddalore district of Tamil Nadu to store water for drinking and irrigation purposes. Vrnam is a 16-kilometre (9.9 mi) long tank with a storage capacity of 1,465,000,000 cubic feet (41,500,000 m).

RAIN WATER HARVESTING TECHNIQUES

There are two main techniques of rain water harvestings.

- Storage of rainwater on surface for future use.
- Recharge to ground water.

The storage of rain water on surface is a traditional techniques and structures used were underground tanks, ponds, check dams, weirs etc. Recharge to ground water is a new concept of rain water harvesting and the structures generally used are :-

Pits :- Recharge pits are constructed for recharging the shallow aquifer. These are constructed 1 to 2 m, wide and to 3 m. deep which are back filled with boulders, gravels, coarse sand.

Trenches:- These are constructed when the permeable stram is available at shallow depth. Trench may be 0.5 to 1 m. wide, 1 to 1.5m. deep and 10 to 20 m. long depending up availability of water. These are back filled with filter. materials.

Dug wells:- Existing dug wells may be utilised as recharge structure and water should pass through filter media before putting into dug well.

Hand pumps :- The existing hand pumps may be used for recharging the shallow/deep aquifers, if the availability of water is limited. Water should pass through filter media before diverting it into hand pumps.

Recharge wells :- Recharge wells of 100 to 300 mm. diameter are generally constructed for recharging the deeper aquifers and water is passed through filter media to avoid choking of recharge wells.

Recharge Shafts :- For recharging the shallow aquifer which are located below clayey surface, recharge shafts of 0.5 to 3 m. diameter and 10 to 15 m. deep are constructed and back filled with boulders, gravels & coarse sand.

Lateral shafts with bore wells :- For recharging the upper as well as deeper aquifers lateral shafts of 1.5 to 2 m. wide & 10 to 30 m. long depending upon availability of water with one or two bore wells are constructed. The lateral shafts is back filled with boulders, gravels & coarse sand. Spreading techniques :- When permeable strata starts from top then this technique is used. Spread the water in streams/Nalas by making check dams, nala bunds, cement plugs, gabion structures or a percolation pond may be constructed.

Diversion of Run off into Existing Surface Water Bodies

Construction activity in and around the city is resulting in the drying up of water bodies and reclamation of these tanks for conversion into plots for houses.

Free flow of storm run off into these tanks and water bodies must be ensured. The storm run off may be diverted into the nearest tanks or depression, which will create additional recharge.

Urbanisation effects on Groundwater Hydrology :

- Increase in water demand
- More dependence on ground water use
- Over exploitation of ground water
- Increase in run-off, decline in well yields and fall in water levels
- Reduction in open soil surface area
- Reduction in infiltration and deterioration in water quality

Methods of artificial recharge in urban areas :

- Water spreading
- Recharge through pits, trenches, wells, shafts
- Rooftop collection of rainwater

- Roadtop collection of rainwater
- Induced recharge from surface water bodies.

Computation of artificial recharge from Roof top rainwater collection :

- Factors taken for computation .
- Roof top area 100 sq.m. for individual house and 500 sq.m. for multi-storied building.
- Average annual monsoon rainfall - 780 mm.
- Effective annual rainfall contributing to recharge 70% - 550 mm.

Benefits of Artificial Recharge in Urban Areas :

- Improvement in infiltration and reduction in run-off.
- Improvement in groundwater levels and yields.
- Reduces strain on Special Village Panchayats/ Municipal / Municipal Corporation water supply
- Improvement in groundwater quality
- Estimated quantity of additional recharge from 100 sq. m. roof top area is 55.000 liters.

HARVESTING RAINWATER HARNESSING LIFE

A Noble Goal - A Common Responsibility

Ground water exploitation is inevitable is Urban areas. But the groundwater potential is getting reduced due to urbanisation resulting in over exploitation. Hence, a strategy to implement the groundwater recharge, in a major way need to be launched with concerted efforts by various Governmental and Non-Governmental Agencies and Public at large to build up the water table and make the groundwater resource, a reliable and sustainable source for supplementing water supply needs of the urban dwellers. Recharge of groundwater through storm run-off and roof top water collection, diversion and collection of run-off into dry tanks, play grounds, parks and other vacant places are to be implemented by Special Village Panchayats/ Municipalities /Municipal Corporations and other Government Establishments with special efforts.

The Special Village Panchayats /Municipalities/Municipal Corporations will help the citizens and builders to adopt suitable recharge method in one's own house or building through demonstration and offering subsidies for materials and incentives, if possible.

Components of a Rainwater Harvesting System

A rainwater harvesting system comprises components of various stages - transporting rainwater through pipes or drains, filtration, and storage in tanks for reuse or recharge. The common components of a rainwater harvesting system involved in these stages are illustrated here.

Catchments

The catchment of a water harvesting system is the surface which directly receives the rainfall and provides water to the system. It can be a paved area like a terrace or courtyard of a building, or an unpaved area like a lawn or open ground. A roof made of reinforced cement concrete (RCC), galvanised iron or corrugated sheets can also be used for water harvesting.

- Rooftop catchments: In the most basic form of this technology, rainwater is collected in simple vessels at the edge of the roof. Variations on this basic approach include collection of rainwater in gutters which drain to the collection vessel through down-pipes constructed for this purpose, and/or the diversion of rainwater from the gutters to containers for settling particulates before being conveyed to the storage container for the domestic use. As the rooftop is the main catchment area, the amount and quality of rainwater collected depends on the area and type of roofing material. Reasonably pure rainwater can be collected from roofs constructed with galvanized corrugated iron, aluminium or asbestos cement sheets, tiles and slates, although thatched roofs tied with bamboo gutters and laid in proper slopes can produce almost the same amount of runoff less expensively (Gould, 1992). However, the bamboo roofs are least suitable because of possible health hazards. Similarly, roofs with metallic paint or other coatings are not recommended as they may impart tastes or colour to the collected water. Roof catchments should also be cleaned regularly to remove dust, leaves and bird droppings so as to maintain the quality of the product water.

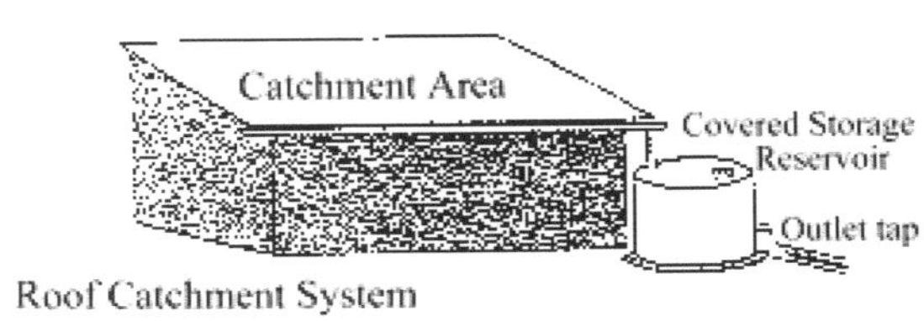

Figure 1: Rooftop Catchment System.

- Land surface catchments: Rainwater harvesting using ground or land surface catchment areas is less complex way of collecting rainwater. It involves improving runoff capacity of the land surface through various techniques including collection of runoff with drain pipes and storage of collected water. Compared to rooftop catchment techniques, ground catchment techniques provide more opportunity for collecting water from a larger surface area. By retaining the flows (including flood flows) of small creeks and streams in small storage reservoirs (on surface or underground) created by low cost (e.g., earthen) dams, this technology can meet water demands during dry periods. There is a possibility of high rates of water loss due to infiltration into the ground, and, because of the often marginal quality of the water collected, this technique is mainly suitable for storing water for agricultural purposes. Various techniques available for increasing the runoff within ground catchment areas involve: i) clearing or altering vegetation cover, ii) increasing the land slope with artificial ground cover, and iii) reducing soil permeability by the soil compaction and application of chemicals (see figure 2).

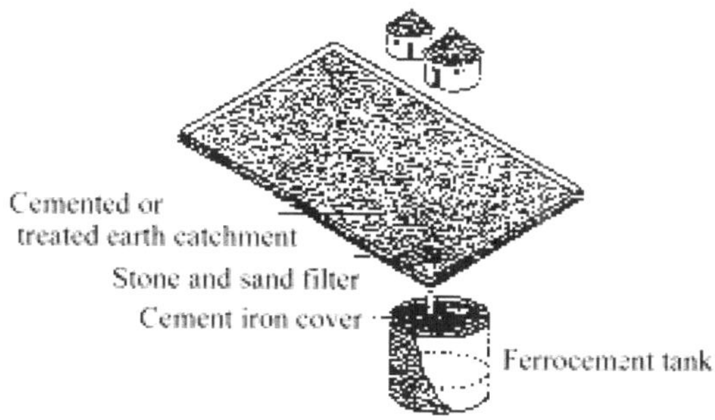

Figure 2: Ground Catchment System

- Clearing or altering vegetation cover: Clearing vegetation from the ground can increase surface runoff but also can induce more soil erosion. Use of dense vegetation cover such as grass is usually suggested as it helps to both maintain an high rate of runoff and minimize soil erosion.
- Increasing slope: Steeper slopes can allow rapid runoff of rainfall to the collector. However, the rate of runoff has to be controlled to minimise soil erosion from the catchment field. Use of plastic sheets, asphalt or tiles along with slope can further increase efficiency by reducing both

evaporative losses and soil erosion. The use of flat sheets of galvanized iron with timber frames to prevent corrosion was recommended and constructed in the State of Victoria, Australia, about 65 years ago (Kenyon, 1929; cited in UNEP, 1982).

- Soil compaction by physical means: This involves smoothing and compacting of soil surface using equipment such as graders and rollers. To increase the surface runoff and minimize soil erosion rates, conservation bench terraces are constructed along a slope perpendicular to runoff flow. The bench terraces are separated by the sloping collectors and provision is made for distributing the runoff evenly across the field strips as sheet flow. Excess flows are routed to a lower collector and stored (UNEP, 1982).
- Soil compaction by chemical treatments: In addition to clearing, shaping and compacting a catchment area, chemical applications with such soil treatments as sodium can significantly reduce the soil permeability. Use of aqueous solutions of a silicone-water repellent is another technique for enhancing soil compaction technologies. Though soil permeability can be reduced through chemical treatments, soil compaction can induce greater rates of soil erosion and may be expensive. Use of sodium-based chemicals may increase the salt content in the collected water, which may not be suitable both for drinking and irrigation purposes.

COLLECTION DEVICES

- Storage tanks: Storage tanks for collecting rainwater harvested using guttering may be either above or below the ground. Precautions required in the use of storage tanks include provision of an adequate enclosure to minimise contamination from human, animal or other environmental contaminants, and a tight cover to prevent algal growth and the breeding of mosquitos. Open containers are not recommended for collecting water for drinking purposes. Various types of rainwater storage facilities can be found in practice. Among them are cylindrical ferrocement tanks and mortar jars. The ferrocement tank consists of a lightly reinforced concrete base on which is erected a circular vertical cylinder with a 10 mm steel base. This cylinder is further wrapped in two layers of light wire mesh to form the frame of the tank. Mortar jars are large jar shaped vessels constructed from wire reinforced mortar. The storage capacity needed should be calculated to take into consideration the length of any dry spells, the amount of rainfall, and the per capita water consumption rate. In most of the Asian countries, the winter months are dry, sometimes for weeks on end, and the annual average rainfall can occur within just a few days. In such circumstances, the storage capacity should be large enough to cover the demands of two to three weeks. For example, a three person household should have a minimum capacity of 3 (Persons) x 90 (l) x 20 (days) = 5 400 l.

- Rainfall water containers: As an alternative to storage tanks, battery tanks (i.e., interconnected tanks) made of pottery, ferrocement, or polyethylene may be suitable. The polyethylene tanks are compact but have a large storage capacity (ca. 1 000 to 2 000 l), are easy to clean and have many openings which can be fitted with fittings for connecting pipes. In Asia, jars made of earthen materials or ferrocement tanks are commonly used. During the 1980s, the use of rainwater catchment technologies, especially roof catchment systems, expanded rapidly in a number of regions, including Thailand where more than ten million 2 m3 ferrocement rainwater jars were built and many tens of thousands of larger ferrocement tanks were constructed between 1991 and 1993. Early problems with the jar design were quickly addressed by including a metal cover using readily available, standard brass fixtures. The immense success of the jar programme springs from the fact that the technology met a real need, was affordable, and invited community participation. The programme also captured the imagination and support of not only the citizens, but also of government at both local and national levels as well as community based organizations, small-scale enterprises and donor agencies. The introduction and rapid promotion of Bamboo reinforced tanks, however, was less successful because the bamboo was attacked by termites, bacteria and fungus. More than 50 000 tanks were built between 1986 and 1993 (mainly in Thailand and Indonesia) before a number started to fail, and, by the late 1980s, the bamboo reinforced tank design, which had promised to provide an excellent low-cost alternative to ferrocement tanks, had to be abandoned.

Gutters

Channels all around the edge of a sloping roof to collect and transport rainwater to the storage tank. Gutters can be semi-circular or rectangular and could be made using:

- Locally available material such as plain galvanised iron sheet (20 to 22 gauge), folded to required shapes.

- Semi-circular gutters of PVC material can be readily prepared by cutting those pipes into two equal semi-circular channels.
- Bamboo or betel trunks cut vertically in half.

The size of the gutter should be according to the flow during the highest intensity rain. It is advisable to make them 10 to 15 per cent oversize.

Gutters need to be supported so they do not sag or fall off when loaded with water. The way in which gutters are fixed depends on the construction of the house; it is possible to fix iron or timber brackets into the walls, but for houses having wider eaves, some method of attachment to the rafters is necessary.

Conduits

Conduits are pipelines or drains that carry rainwater from the catchment or rooftop area to the harvesting system. Conduits can be of any material like polyvinyl chloride (PVC) or galvanized iron (GI), materials that are commonly available.

CONVEYANCE SYSTEMS

Conveyance systems are required to transfer the rainwater collected on the rooftops to the storage tanks. This is usually accomplished by making connections to one or more down-pipes connected to the rooftop gutters. When selecting a conveyance system, consideration should be given to the fact that, when it first starts to rain, dirt and debris from the rooftop and gutters will be washed into the down-pipe. Thus, the relatively clean water will only be available some time later in the storm. There are several possible choices to selectively collect clean water for the storage tanks. The most common is the down-pipe flap. With this flap it is possible to direct the first flush of water flow through the down-pipe, while later rainfall is diverted into a storage tank. When it starts to rain, the flap is left in the closed position, directing water to the down-pipe, and, later, opened when relatively clean water can be collected. A great disadvantage of using this type of conveyance control system is the necessity to observe the runoff quality and manually operate the flap. An alternative approach would be to automate the opening of the flap as described below.

A funnel-shaped insert is integrated into the down-pipe system. Because the upper edge of the funnel is not in direct contact with the sides of the down-pipe, and a small gap exists between the down-pipe walls and the funnel, water is free to flow both around the funnel and through the funnel. When it first starts to rain, the volume of water passing down the pipe is small, and the *dirty* water runs down the walls of the pipe, around the funnel and is discharged to the ground as is normally the case with rainwater guttering. However, as the rainfall continues, the volume of water increases and *clean* water fills the down-pipe. At this higher volume, the funnel collects the clean water and redirects it to a storage tank. The pipes used for the collection of rainwater, wherever possible, should be made of plastic, PVC or other inert substance, as the pH of rainwater can be low (acidic) and could cause corrosion, and mobilization of metals, in metal pipes.

In order to safely fill a rainwater storage tank, it is necessary to make sure that excess water can overflow, and that blockages in the pipes or dirt in the water do not cause damage or contamination of the water supply. The design of the funnel system, with the drain-pipe being larger than the rainwater tank feed-pipe, helps to ensure that the water supply is protected by allowing excess water to bypass the storage tank. A modification of this design is shown in Figure 5, which illustrates a simple overflow/bypass system. In this system, it also is possible to fill the tank from a municipal drinking water source, so that even during a prolonged drought the tank can be kept full. Care should be taken, however, to ensure that rainwater does not enter the drinking water distribution system.

Transportation

Rainwater from rooftop should be carried through down take water pipes or drains to storage/harvesting system. Water pipes should be UV resistant (ISI HDPE/PVC pipes) of required capacity. Water from sloping roofs could be caught through gutters and down take pipe. At terraces, mouth of the each drain should have wire mesh to restrict floating material.

First-flushing

A first flush device is a valve that ensures that runoff from the first spell of rain is flushed out and does not enter the system. This needs to be done since the first spell of rain carries a relatively larger amount of pollutants from the air and catchment surface.

First flush is a device used to flush off the water received in first shower. The first shower of rains needs to be flushed-off to avoid contaminating storable/ rechargeable water by the probable contaminants of the atmosphere and the catchment roof. It will also help in cleaning of silt and other material deposited on roof during dry seasons Provisions of first rain separator should be made at outlet of each drainpipe.

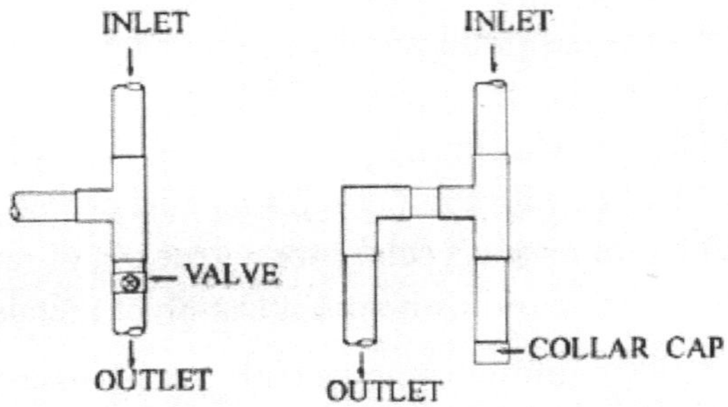

Filter

There is always some skepticism regarding Roof Top Rainwater Harvesting since doubts are raised that rainwater may contaminate groundwater. There is remote possibility of this fear coming true if proper filter mechanism is not adopted. Secondly all care must be taken to see that underground sewer drains are not punctured and no leakage is taking place in close vicinity. Filters are used fro treatment of water to effectively remove turbidity, colour and microorganisms. After first flushing of rainfall, water should pass through filters. There are different types of filters in practice, but basic function is to purify water.

SAND GRAVEL FILTER

These are commonly used filters, constructed by brick masonry and filleted by pebbles, gravel, and sand as shown in the figure. Each layer should be separated by wire mesh.

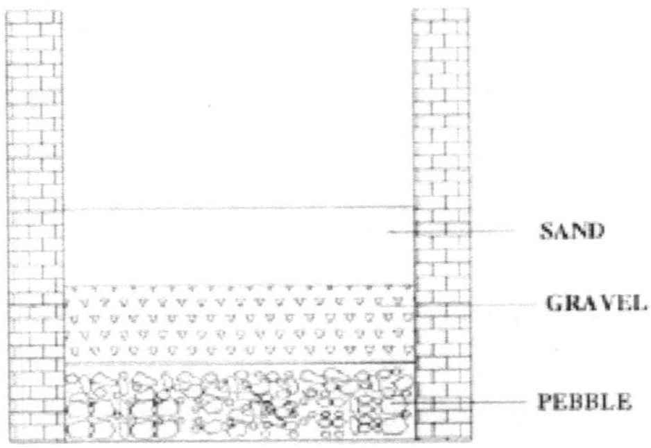

SAND GRAVEL FILTER

Charcoal Filter

Charcoal filter can be made in-situ or in a drum. Pebbles, gravel, sand and charcoal as shown in the figure should fill the drum or chamber. Each layer should be separated by wire mesh. Thin layer of charcoal is used to absorb odor if any.

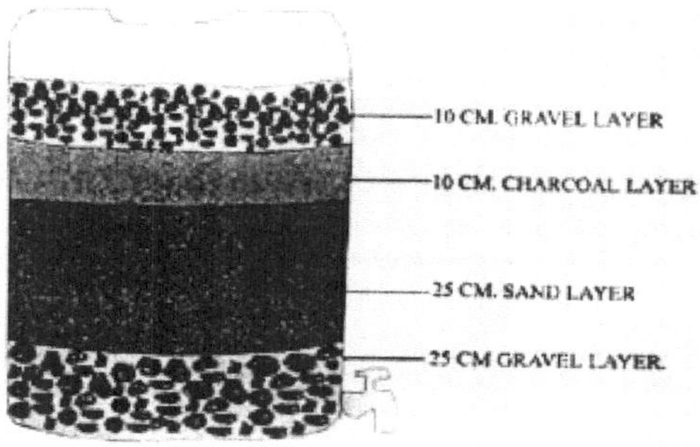

PVC- Pipe filter

This filter can be made by PVC pipe of 1 to 1.20 m length; Diameter of pipe depends on the area of roof. Six inches dia. pipe is enough for a 1500 Sq. Ft. roof and 8 inches dia. pipe should be used for roofs more then 1500 Sq. Ft. Pipe is divided into three compartments by wire mesh. Each component should be filled with gravel and sand alternatively as shown in the figure. A layer of charcoal could also be inserted between two layers. Both ends of filter should have reduce of required size to connect inlet and outlet. This filter could be placed horizontally or vertically in the system.

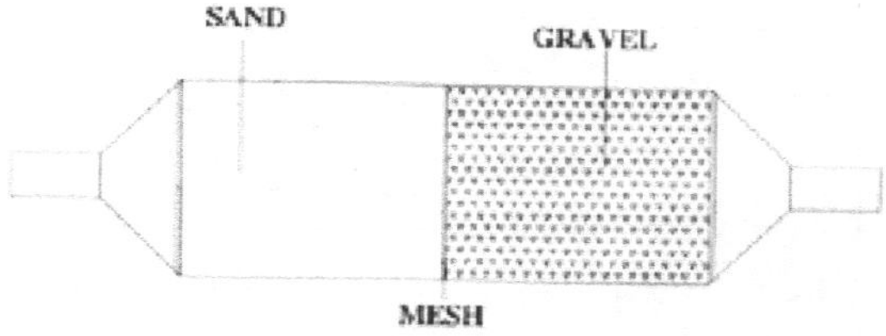

PVC PIPE FILTER

Sponge Filter

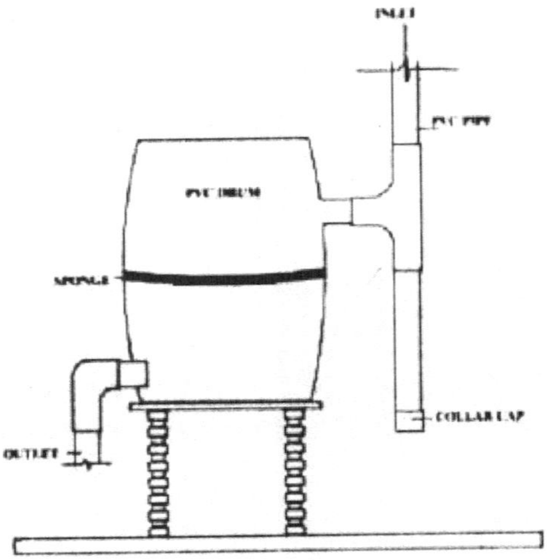

SPONGE FILTER

It is a simple filter made from PVC drum having a layer of sponge in the middle of drum. It is the easiest and cheapest form filter, suitable for residential units.

METHODS OF ROOF TOP RAINWATER HARVESTING

Storage of Direct use

In this method rain water collected from the roof of the building is diverted to a storage tank. The storage tank has to be designed according to the water requirements, rainfall and catchment availability. Each drainpipe should have mesh filter at mouth and first flush device followed by filtration system before connecting to the storage tank. It is advisable that each tank should have excess water over flow system.

Excess water could be diverted to recharge system. Water from storage tank can be used for secondary purposes such as washing and gardening etc. This is the most cost effective way of rainwater harvesting. The main advantage of collecting and using the rainwater during rainy season is not only to save water from conventional sources, but also to save energy incurred on transportation and distribution of water at the doorstep. This also conserve groundwater, if it is being extracted to meet the demand when rains are on.

Recharging ground water aquifers

Ground water aquifers can be recharged by various kinds of structures to ensure percolation of rainwater in the ground instead of draining away from the surface. Commonly used recharging methods are:

(a) Recharging of bore wells

(b) Recharging of dug wells.

(c) Recharge pits

(d) Recharge Trenches

(e) Soak ways or Recharge Shafts

(f) Percolation Tanks

RECHARGING OF BORE WELLS

Rainwater collected from rooftop of the building is diverted through drainpipes to settlement or filtration tank. After settlement filtered water is diverted to bore wells to recharge deep aquifers. Abandoned bore wells can also be used for recharge. Optimum capacity of settlement tank/filtration tank can be designed on the basis of area of catchement, intensity of rainfall and recharge rate as discussed in design parameters. While recharging, entry of floating matter and silt should be restricted because it may clog the recharge structure. "first one or two shower should be flushed out through rain separator to avoid contamination. This is very important, and all care should be taken to ensure that this has been done."

RECHARGE PITS

Recharge pits are small pits of any shape rectangular, square or circular, contracted with brick or stone masonry wall with weep hole at regular intervals. to of pit can be covered with perforated covers. Bottom of pit should be filled with filter media.

The capacity of the pit can be designed on the basis of catchment area, rainfall intensity and recharge rate of soil. Usually the dimensions of the pit may be of 1 to 2 m width and 2 to 3 m deep depending on the depth of pervious strata. These pits are suitable for recharging of shallow aquifers, and small houses.

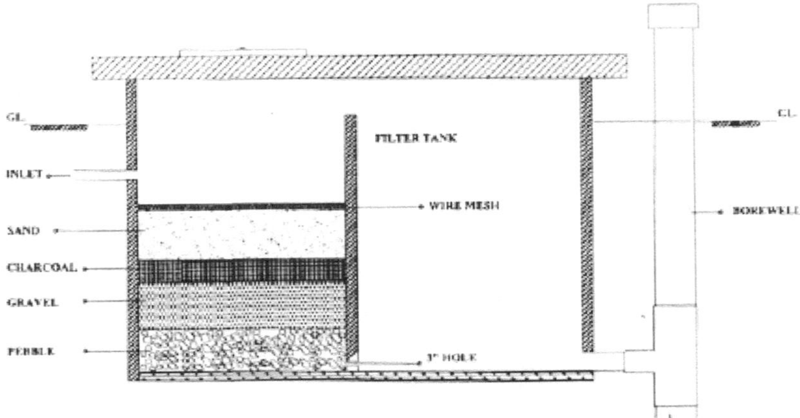

FILTRATION TANK

SOAK AWAY OR RECHARGE SHAFTS

Soak away or recharge shafts are provided where upper layer of soil is alluvial or less pervious. These are bored hole of 30 cm dia. up to 10 to 15 m deep, depending on depth of pervious layer. Bore should be lined with slotted/ perforated PVC/MS pipe to prevent collapse of the vertical sides. At the top of soak away required size sump is constructed to retain runoff before the filters through soak away. Sump should be filled with filter media.

Recharging of dug wells

Dug well can be used as recharge structure. Rainwater from the rooftop is diverted to dug wells after passing it through filtration bed. Cleaning and desalting of dug well should be done regularly to enhance the recharge rate. The filtration method suggested for bore well recharging could be used.

Recharge Trenches

Recharge trench in provided where upper impervious layer of soil is shallow. It is a trench excavated on the ground and refilled with porous media like pebbles, boulder or brickbats. it is usually made for harvesting the surface runoff. Bore wells can also be provided inside the trench as recharge shafts to enhance percolation. The length of the trench is decided as per the amount of runoff expected. This method is suitable for small houses, playgrounds, parks and roadside drains. The recharge trench can be of size 0.50 to 1.0 m wide and 1.0 to 1.5 m deep.

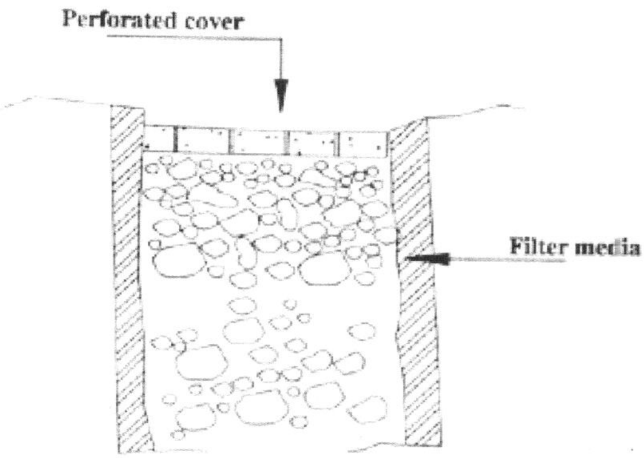

Recharge Trenches

Percolation tanks

Percolation tanks are artificially created surface water bodies, submerging a land area with adequate permeability to facilitate sufficient percolation to recharge the ground water. These can be built in big campuses where land is available and topography is suitable.

Surface run-off and roof top water can be diverted to this tank. Water accumulating in the tank percolates in the solid to augment the ground water. The stored water can be used directly for gardening and raw use. Percolation tanks should be built in gardens, open spaces and roadside green belts of urban area.

Do's and Don'ts

Harvested rainwater is used for direct usage or for recharging aquifers. It is most important to ensure that the rainwater caught is free from pollutants. Following precautionary measures should be taken while harvesting rainwater:

- Roof or terraces uses for harvesting should be clean, free from dust, algal plants etc.
- Roof should not be painted since most paints contain toxic substances and may peel off.
- Do not store chemicals, rusting iron, manure or detergent on the roof.
- Nesting of birds on the roof should be prevented.
- Terraces should not be used for toilets either by human beings or by pets.
- Provide gratings at mouth of each drainpipe on terraces to trap leaves debris and floating materials.

- Provision of first rain separator should be made to flush off first rains.
- Do not use polluted water to recharge ground water.
- Ground water should only be recharged by rainwater.
- Before recharging, suitable arrangements of filtering should be provided.
- Filter media should be cleaned before every monsoon season.
- During rainy season, the whole system (roof catchment, pipes, screens, first flush, filters, tanks) should be checked before and after each rain and preferably cleaned after every dry period exceeding a month.
- At the end of the dry season and just before the first shower of rain is anticipated, the storage tank should be scrubbed and flushed off all sediments and debris.

5

WATER CONSERVATION

Water conservationist encompasses the policies, strategies and activities to manage fresh water as a sustainable resource, to protect the water environment, and to meet current and future human demand. Population, household size and growth and affluence all affect how much water is used. Factors such as climate change will increase pressures on natural water resources especially in manufacturing and agricultural irrigation.

The goals of water conservation efforts include:

- Ensuring availability of water for future generations. This requires that the withdrawal of fresh water from an ecosystem does not exceed its natural replacement rate.
- Energy conservation. Water pumping, delivery and waste water treatment facilities consume a significant amount of energy. In some regions of the world over 15% of total electricity consumption is devoted to water management.
- Habitat conservation. Minimizing human water use helps to preserve fresh fire habitats for local wildlife and migrating waterfowl, as well as reducing the need to build new dams and other water diversion infrastructures.

Strategies

In implementing water conservation principles, there are a number of key activities that may be beneficial.

1. Any beneficial reduction in water loss, use and waste of resources.
2. Avoiding any damage to water quality.
3. Improving water management practices that reduce or enhance the beneficial use of water.

Social solutions

Water conservation programs involved in social solutions are typically initiated at the local level, by either municipal water utilities or regional governments. Common strategies include public outreach campaigns, tiered water rates (charging progressively higher prices as water use increases), or restrictions on outdoor water use such as lawn watering and car washing. Cities in dry climates often require or encourage the installation of xeriscaping or natural landscaping in new homes to reduce outdoor water usage.

One fundamental conservation goal is universal metering. The prevalence of residential water metering varies significantly worldwide. Recent studies have estimated that water supplies are metered in less than 30% of UK households, and about 61% of urban Canadian homes (as of 2001). Although individual water meters have often been considered impractical in homes with private wells or in multifamily buildings, the U.S. Environmental Protection Agency estimates that metering alone can reduce consumption by 20 to 40 percent. In addition to raising consumer awareness of their water use, metering is also an important way to identify and localize water leakage. Water metering would benefit society in the long run it is proven that water metering increases the efficiency of the entire water system, as well as help unnecessary expenses for individuals for years to come. One would be unable to waste water unless they are willing to pay the extra charges, this way the water department would be able to monitor water usage by public, domestic and manufacturing services.

Some researchers have suggested that water conservation efforts should be primarily directed at farmers, in light of the fact that crop irrigation accounts for 70% of the world's fresh water use. The agricultural sector of most countries is important both economically and politically, and water subsidies are common. Conservation advocates have urged removal of all subsidies to force farmers to grow more water-efficient crops and adopt less wasteful irrigation techniques.

New technology poses a few new options for consumers, features such and full flush and half flush when using a toilet are trying to make a difference in water consumption and waste. Also available in our modern world is shower heads that help reduce wasting water, old shower heads are said to use 5-10 gallons per minute. All new fixtures available are said to use 2.5 gallons per minute and offer equal water coverage.

Household applications

The Home Water Works website contains useful information on household water conservation. Contrary to popular view, experts suggest the most efficient way is replacing toilets and retrofitting washers.

Water-saving technology for the home includes:

1. Low-flow shower heads sometimes called energy-efficient shower heads as they also use less energy
2. Low-flush toilets and composting toilets. These have a dramatic impact in the developed world, as conventional Western toilets use large volumes of water
3. Dual flush toilets created by C aroma includes two buttons or handles to flush different levels of water. Dual flush toilets use up to 67% less water than conventional toilets
4. Faucet aerators, which break water flow into fine droplets to maintain "wetting effectiveness" while using less water. An additional benefit is that they reduce splashing while washing hands and dishes
5. Raw water flushing where toilets use sea water or non-purified water
6. Waste water reuse or recycling systems, allowing:
   - o Reuse of graywater for flushing toilets or watering gardens
   - o Recycling of wastewater through purification at a water treatment plant.
7. Rainwater harvesting
8. High-efficiency clothes washers
9. Weather-based irrigation controllers
10. Garden hose nozzles that shut off water when it is not being used, instead of letting a hose run.
11. Low flow taps in wash basins
12. Swimming pool covers that reduce evaporation and can warm pool water to reduce water, energy and chemical costs.
13. Automatic faucet is a water conservation faucet that eliminates water waste at the faucet. It automates the use of faucets without the use of hands.

COMMERCIAL APPLICATIONS

Many water-saving devices (such as low-flush toilets) that are useful in homes can also be useful for business water saving. Other water-saving technology for businesses includes:

- Waterless urinals
- Waterless car washes
- Infrared or foot-operated taps, which can save water by using short bursts of water for rinsing in a kitchen or bathroom
- Pressurized waterbrooms, which can be used instead of a hose to clean sidewalks
- X-ray film processor re-circulation systems
- Cooling tower conductivity controllers
- Water-saving steam sterilizers, for use in hospitals and health care facilities
- Rain water harvesting
- Water to Water heat exchangers.

AGRICULTURAL APPLICATIONS

For crop irrigation, optimal water efficiency means minimizing losses due to evaporation, runoff or subsurface drainage while maximizing production. An evaporation pan in combination with specific crop correction factors can be used to determine how much water is needed to satisfy plant requirements. Flood irrigation, the oldest and most common type, is often very uneven in distribution, as parts of a field may receive excess water in order to deliver sufficient quantities to other parts. Overhead irrigation, using center-pivot or lateral-moving sprinklers, has the potential for a much more equal and controlled distribution pattern. Drip irrigation is the most expensive and least-used type, but offers the ability to deliver water to plant roots with minimal losses. However, drip irrigation is increasingly affordable, especially for the home gardener and in light of rising water rates. There are also cheap effective methods similar to drip irrigation such as the use of soaking hoses that can even be submerged in the growing medium to eliminate evaporation.

As changing irrigation systems can be a costly undertaking, conservation efforts often concentrate on maximizing the efficiency of the existing system. This may include chiseling compacted soils, creating furrow dikes to prevent runoff, and using soil moisture and rainfall sensors to optimize irrigation schedules. Usually large gains in efficiency are possible through measurement and more effective management of the existing irrigation system. The 2011 UNEP Green Economy Report notes that "[i]mproved soil organic matter from the use of green manures, mulching, and recycling of crop residues and animal manure increases the water holding capacity of soils and their ability to absorb water during torrential rains", which is a way to optimize the use of rainfall and irrigation during dry periods in the season.

Our ancient religious texts and epics give a good insight into the water storage and conservation systems that prevailed in those days.

Over the years rising populations, growing industrialization, and expanding agriculture have pushed up the demand for water. Efforts have been made to collect water by building dams and reservoirs and digging wells; some countries have also tried to recycle and desalinate (remove salts) water. Water conservation has become the need of the day. The idea of ground water recharging by harvesting rainwater is gaining importance in many cities.

In the forests, water seeps gently into the ground as vegetation breaks the fall. This groundwater in turn feeds wells, lakes, and rivers. Protecting forests means protecting water 'catchments'. In ancient India, people believed that forests were the 'mothers' of rivers and worshipped the sources of these water bodies.

Some ancient Indian methods of water conservation

The Indus Valley Civilization, that flourished along the banks of the river Indus and other parts of western and northern India about 5,000 years ago, had one of the most sophisticated urban water supply and sewage systems in the world. The fact that the people were well acquainted with hygiene can be seen from the covered drains running beneath the streets of the ruins at both Mohenjodaro and Harappa. Another very good example is the well-planned city of Dholavira, on Khadir Bet, a low plateau in the Rann in Gujarat. One of the oldest water harvesting systems is found about 130 km from Pune along Naneghat in the Western Ghats. A large number of tanks were cut in the rocks to provide drinking water to tradesmen who used to travel along this ancient trade route. Each fort in the area had its own water harvesting and storage system in the form of rock-cut cisterns, ponds, tanks and wells that are still in use today. A large number of forts like Raigad had tanks that supplied water.

- In ancient times, houses in parts of western Rajasthan were built so that each had a rooftop water harvesting system. Rainwater from these rooftops was directed into underground tanks. This system can be seen even today in all the forts, palaces and houses of the region.

- Underground baked earthen pipes and tunnels to maintain the flow of water and to transport it to distant places, are still functional at Burhanpur in Madhya Pradesh, Golkunda and Bijapur in Karnataka, and Aurangabad in Maharashtra.

RAINWATER HARVESTING

In urban areas, the construction of houses, footpaths and roads has left little exposed earth for water to soak in. In parts of the rural areas of India, floodwater quickly flows to the rivers, which then dry up soon after the rains stop. If this water can be held back, it can seep into the ground and recharge the groundwater supply.

This has become a very popular method of conserving water especially in the urban areas. Rainwater harvesting essentially means collecting rainwater on the roofs of building and storing it underground for later use. Not only does this recharging arrest groundwater depletion, it also raises the declining water table and can help augment water supply. Rainwater harvesting and artificial recharging are becoming very important issues. It is essential to stop the decline in groundwater levels, arrest sea-water ingress, i.e. prevent sea-water from moving landward, and conserve surface water run-off during the rainy season.

Town planners and civic authority in many cities in India are introducing bylaws making rainwater harvesting compulsory in all new structures. No water or sewage connection would be given if a new building did not have provisions for rainwater harvesting. Such rules should also be implemented in all the other cities to ensure a rise in the groundwater level.

Realizing the importance of recharging groundwater, the CGWB (Central Ground Water Board) is taking steps to encourage it through rainwater harvesting in the capital and elsewhere. A number of government buildings have been asked to go in for water harvesting in Delhi and other cities of India.

All you need for a water harvesting system is rain, and a place to collect it! Typically, rain is collected on rooftops and other surfaces, and the water is carried down to where it can be used immediately or stored. You can direct water run-off from this surface to plants, trees or lawns or even to the aquifer.

Some of the benefits of rainwater harvesting are as follows

- Increases water availability

- Checks the declining water table
- Is environmentally friendly
- Improves the quality of groundwater through the dilution of fluoride, nitrate, and salinity
- Prevents soil erosion and flooding especially in urban areas

Agriculture

Conservation of water in the agricultural sector is essential since water is necessary for the growth of plants and crops. A depleting water table and a rise in salinity due to overuse of chemical fertilizers and pesticides has made matters serious. Various methods of water harvesting and recharging have been and are being applied all over the world to tackle the problem. In areas where rainfall is low and water is scarce, the local people have used simple techniques that are suited to their region and reduce the demand for water.

In India's arid and semi-arid areas, the 'tank' system is traditionally the backbone of agricultural production. Tanks are constructed either by bunding or by excavating the ground and collecting rainwater.

Rajasthan, located in the Great Indian Desert, receives hardly any rainfall, but people have adapted to the harsh conditions by collecting whatever rain falls. Large bunds to create reservoirs known as khadin, dams called johads, tanks, and other methods were applied to check water flow and accumulate run-off. At the end of the monsoon season, water from these structures was used to cultivate crops. Similar systems were developed in other parts of the country. These are known by various local names ¾ jal talais in Uttar Pradesh, the haveli system in Madhya Pradesh, ahar in Bihar, and so on.

Reducing water demand

Simple techniques can be used to reduce the demand for water. The underlying principle is that only part of the rainfall or irrigation water is taken up by plants, the rest percolates into the deep groundwater, or is lost by evaporation from the surface. Therefore, by improving the efficiency of water use, and by reducing its loss due to evaporation, we can reduce water demand.

There are numerous methods to reduce such losses and to improve soil moisture. Some of them are listed below.

Mulching, i.e., the application of organic or inorganic material such as plant debris, compost, etc., slows down the surface run-off, improves the soil moisture, reduces evaporation losses and improves soil fertility.

Soil covered by crops, slows down run-off and minimizes evaporation losses. Hence, fields should not be left bare for long periods of time. Ploughing helps to move the soil around. As a consequence it retains more water thereby reducing evaporation.

Shelter belts of trees and bushes along the edge of agricultural fields slow down the wind speed and reduce evaporation and erosion.

Planting of trees, grass, and bushes breaks the force of rain and helps rainwater penetrate the soil.

Fog and dew contain substantial amounts of water that can be used directly by adapted plant species. Artificial surfaces such as netting-surfaced traps or polyethylene sheets can be exposed to fog and dew. The resulting water can be used for crops.

Contour farming is adopted in hilly areas and in lowland areas for paddy fields. Farmers recognize the efficiency of contour-based systems for conserving soil and water. Salt-resistant varieties of crops have also been developed recently. Because these grow in saline areas, overall agricultural productivity is increased without making additional demands on freshwater sources. Thus, this is a good water conservation strategy.

Transfer of water from surplus areas to deficit areas by inter-linking water systems through canals, etc.

Desalination technologies such as distillation, electro-dialysis and reverse osmosis are available.

Use of efficient watering systems such as drip irrigation and sprinklers will reduce the water consumption by plants.

Water conservation

The most important step in the direction of finding solutions to issues of water and environmental conservation is to change people's attitudes and habits¾this includes each one of us. Conserve water because it is the right thing to do. We can follow some of the simple things that have been listed below and contribute to water conservation. Try to do one thing each day that will result in saving water. Don't worry if the savings are minimal¾every drop counts! You can make a difference.

Remember to use only the amount you actually need.

Form a group of water-conscious people and encourage your friends and neighbours to be part of this group. Promote water conservation in community newsletters and on bulletin boards. Encourage your friends, neighbours and co-workers to also contribute.

Encourage your family to keep looking for new ways to conserve water in and around your home.

Make sure that your home is leak-free. Many homes have leaking pipes that go unnoticed.

Do not leave the tap running while you are brushing your teeth or soaping your face. See that there are no leaks in the toilet tank. You can check this by adding colour to the tank. If there is a leak, colour will appear in the toilet bowl within 30 minutes. (Flush as soon as the test is done, since food colouring may stain the tank.)

Avoid flushing the toilet unnecessarily. Put a brick or any other device that occupies space to cut down on the amount of water needed for each flush. When washing the car, use water from a bucket and not a hosepipe. Do not throw away water that has been used for washing vegetables, rice or dals¾use it to water plants or to clean the floors, etc

You can store water in a variety of ways. A simple method is to place a drum on a raised platform directly under the rainwater collection source. You can also collect water in a bucket during the rainy season.

Why Conserve Water?

There are many, many ways that we use our water and that is partly why it is so important that we conserve our water. Water is our most precious resource. Water is vital to life. Humans, plants, and animals are made up of mostly water. All living things would die if it weren't for water. We use water for drinking, washing, cleaning, cooking, and growing our food as well as many, many other things. The average American uses around 150-250 gallons of water daily. Even more water is used by industries to generate electricity, manufacture things, and transport people and goods.

Common household uses consume a lot of water. It may take between 30 and 40 gallons for one bath. The average toilet uses about 5 gallons of water per flush. It takes 20-40 gallons of water for one shower. Washing machines use an average of 25 gallons per load. The kitchen sink takes roughly 20 gallons per day for preparing food and washing dishes. The bathroom sink, used for washing hands, shaving and brushing teeth, requires about 15 gallons per day. These numbers are estimated for the average household in America.

Much of our fresh water is also used outdoors for watering lawns, flower beds, and vegetable gardens, as well as washing cars and filling swimming pools. We must be careful not to pollute the water that we use outside. Many people use chemicals on their lawns and gardens and then water them with pure water. The water will wash the chemicals off of your plants and then run down a storm drain and go straight to the rivers and streams where fish make their homes. This kind of polluted water can kill fish and wildlife. We should always be mindful of whether or not we are polluting the water around us. We need to take care of our water, not abuse it.

Every town and city, whether small or big, uses water. Cities use water for fire fighting, street cleaning, and watering public areas such as parks, grass, trees, shrubs, and flowers. Water is also used to fill public drinking fountains, including those at schools and libraries. All of the different businesses in your community also use water. Think about all the water that is used by restaurants, hospitals, laundries, dry cleaners, golf courses, hotels, car washes, beauty shops, barber shops, gas stations, and health clubs as well as all of the other businesses in town. These all add up to quite a big demand on Washington's water supply. We need to try to think about how many people need to use water and conserve our water so that there is enough to go around.

The amount of water needed to run a farm is tremendous. When we think of water on a farm, we think of watering crops; but the amount of water needed on a dairy farm is a huge amount also. Chickens, pigs, sheep, and all the other animals in a farmyard need drinking water to stay alive. Food must be grown for them to eat, and water is also required in the cooling systems used to keep production meat fresh. Vegetable and grain crops also require water. Water is used in spreading fertilizers, herbicides, and pesticides, which produce a greater crop yield, but can also contaminate the water. Most of the water used on farms is used for irrigation. Studies show that by using drip irrigation, farmers can conserve up to 60% of the water that it would normally take to irrigate their crops using other systems of irrigation.

It takes about 26 gallons of water to produce one ear of corn while it takes about 2000-2500 gallons of water to produce one pound of beef. Around 120 gallons of water is required to produce one egg. About 300 gallons of water is needed to produce one loaf of bread, and it takes about 12,000 gallons of water to grow a bushel of wheat. Believe it or not, about 1,400 gallons of water are used during the final production of one fast-food meal including a burger, fries, and a soft drink.

Hydroelectric plants are the largest users of water.

Hydroelectric plants capture the kinetic energy of falling water to make electricity. They do this with a dam. The dam forces the water level to go up so that the water will have more power when falling. The force of the falling water pressing against the turbines' blades cause them to spin. The spinning turbines transmit the kinetic energy of the falling water to generators. The generators spin when the turbines spin generating electricity that will be transmitted on the power lines to homes and businesses.

Of all the electricity in the world, about 20% is generated by hydropower. About 10% of all the electricity in the United States is provided by hydropower. Hydropower generating prevents a lot of pollution. Hydropower generating is clean and does not leave any waste. Because of the electricity generated by hydropower, the amount of oil and coal needed to produce enough electricity is reduced. It prevents the need to burn about 22 billion gallons of oil or 120 million tons of coal each year.

The amount of electricity that a hydroelectric plant produces depends on two things: how far the water falls and the quantity of water falling. The higher the dam, the further the water falls and the more electric power produced. If the water falls twice as far, there will be twice as much electricity generated. The quantity of water that falls also affects the amount of power produced. The more water that flows through the turbines making them spin, the more electric power produced.

Water is also essential in industry. It is heated and the steam is used to run machinery. Water is used to cool hot metal such as in the production of steel.

Water is also used to cool the air. It is an important element in many products like chemicals, drugs, lotions, shampoos, cosmetics, cleaners, and also beverages. Water is used in processing food and in innumerable factories and industrial processes including the manufacturing of paper. Water used in processing foods and beverages must be absolutely clean, while other industries such as a manufacturing plant may use a lower quality of water. "In the early 1900s, American industry used about 10 to 15 billion gallons of water a day. With the huge growth in industry following World War II, the industrial use of water also grew. By 1980, industry was using about 150-200 billion gallons each day." - Water: A Resource in Crisis by Eileen Lucas

Recreation is another way that we use and enjoy Washington state's water. Many people enjoy fishing, boating, sailing, canoeing, rafting, and swimming, as well as many other recreational activities that depend on water. Most of us enjoy at least one of these activities and would be disappointed if we could never do any of them again. If our water is continually polluted though, then it could become dangerous to go swimming and all of the fish could die so that no one could go fishing either. No one would enjoy boating, sailing, canoeing, or rafting in polluted water either. We need to remember that it is important to guard our waters from becoming polluted.

Transportation is yet another way that we use water. Many people use boats and ferries to commute to and from work every day. People also enjoy going on cruise ships or just going sailing. Water transportation has been a channel for growth in America, but that growth has not come without cost. The destruction caused by shipping includes fuel spills, which coat the water with oil and coal dust, and erosion of the shoreline which causes damage to the fish and other inhabitants of the water.

6

WATER RECYCLE

Reclaimed water or recycled water, is former wastewater (sewage) that is treated to remove solids and impurities, and used in sustainable landscaping irrigation, to recharge groundwater aquifers, to meet commercial and industrial water needs, and for drinking. The purpose of these processes is water conservation and sustainability, rather than discharging the treated water to surface waters such as rivers and oceans. In some cases, recycled water can be used for streamflow augmentation to benefit ecosystems and improve aesthetics. One example of this is along Calera Creek in the City of Pacifica, CA.

The definition of reclaimed water, as defined by Levine and Asano, is "The end product of wastewater reclamation that meets water quality requirements for biodegradable materials, suspended matter and pathogens." Simply stated, reclaimed water is water that is used more than one time before it passes back into the natural water cycle. Scientifically-proven advances in water technology allow communities to reuse water for many different purposes, including industrial, irrigation, and drinking. The water is treated differently depending upon the source and use of the water and how it gets delivered. Cycled repeatedly through the planetary hydrosphere, all water on Earth is recycled water, but the terms "recycled water" or "reclaimed water" typically mean wastewater sent from a home or business through a pipeline system to a treatment facility, where it is treated to a level consistent with its intended use. The water is then routed directly to a recycled water system for uses such as irrigation or industrial cooling.

There are examples of communities that have safely used recycled water for many years. Los Angeles County's sanitation districts have provided treated wastewater for landscape irrigation in parks and golf courses since 1929. The first reclaimed water facility in California was built at San Francisco's Golden Gate Park in 1932. The Irvine Ranch Water District (IRWD) was the first water district in California to receive an unrestricted use permit from the state for its recycled water; such a permit means that water can be used for any purpose except drinking. IRWD maintains one of the largest recycled water systems in the nation with more than 400 miles serving more than 4,500 metered connections. The Irvine Ranch Water District and Orange County Water District in Southern California are established leaders in recycled water. Further, the Orange County Water District, located in Orange County, and in other locations throughout the world such as Singapore, water is given more advanced treatments and is used indirectly for drinking.

In spite of quite simple methods that incorporate the principles of water-sensitive urban design (WSUD) for easy recovery of stormwater runoff, there remains a common perception that reclaimed water must involve sophisticated and technically complex treatment systems, attempting to recover the most complex and degraded types of sewage. As this effort is driven by sustainability factors, this type of implementation should inherently be associated with point source solutions, where it is most economical to achieve the expected outcomes. Harvesting of stormwater or rainwater can be an extremely simple to comparatively complex, as well as energy and chemical intensive, recovery of more contaminated sewage.

What is recycled water?

Recycled water is wastewater that has been purified so it can be used again for new purposes.

We treat wastewater from our sewage treatment plants to produce high quality recycled water that is suitable for a range of non-drinking purposes.

Why use recycled water?

In an uncertain climate, we must be smart about how we use our finite supplies of drinking water. Recycled water is a good alternative to use in situations such as:

- Agriculture irrigation
- Watering public parks and gardens
- Toilet flushing, car washing, garden watering and other household uses

- Recycled water is just one of the alternative water supplies available to us. Grey water, stormwater harvesting and sewer mining are also helping to preserve our drinking water supplies for drinking. You can read more about them on the following page:
- Alternative water sources

Melbourne Water's role

We are a wholesale producer and supplier of recycled water.

Using world-class technologies tailored to our infrastructure and wastewater quality, we produce recycled water from our two sewage treatment plants, the Western Treatment Plant at Werribee and the Eastern Treatment Plant at Bangholme. We then sell recycled water to retail water companies, who distribute it directly to individual users.

We are also leaders in recycled water technology, and support safe and efficient reuse of water. Our aim is to make sure our water supply system is resilient to periods of low rainfall in the future. We are working closely with water retailers and industry stakeholders to investigate and make better use of all available sources of water, and integrated solutions to managing water. Learn more about this holistic approach:

- Integrated water management

SUPPLY AND DEMAND

We have the capacity to produce the largest amount of Class A recycled water of any organisation in Australia. In Victoria, Class A is the highest class of recycled water, which allows it to be used for a range of non-drinking purposes.

However, demand for recycled water is very dependent on climate. During the drought there was much greater use of recycled water compared to wetter periods. The largest consumers of recycled water are farmers and irrigators, so demand is seasonal and heavily dependent on rainfall. However, as new housing estates gain access to dedicated pipes supplying recycled water, demand will become more consistent. We will continue working with the water retailers and the Victorian Government to investigate new uses for recycled water and long-term solutions to secure our water supplies.

Terminology

There is no one-size-fits-all solution to water reuse, but there are many safe and scientifically-proven options that allow communities to sustain their local water supplies. Below are terms scientists and water experts use to describe some of these reclaimed water options:

Reused water is water used more than once or recycled.

Potable water is drinking water.

Potable reuse refers to reused water you can drink.

Nonpotable reuse refers to reused water that is not used for drinking, but is safe to use for irrigation or industrial purposes.

Renewed water refers to reclaimed or recycled water that has been subjected to advanced treatment to make it potable. The term was coined by civil engineer, Eleanor Allen, as a means of distinguishing renewed water from reclaimed or recycled water which is generally not potable and to provide alternate positive terminology to counter the phrase "toilet-to-tap".

De facto, unacknowledged or unplanned potable reuse occurs when water intakes draw raw water supplies downstream from discharges of treated effluent from wastewater treatment plants/water reclamation facilities or resource recovery facilities. For example, if you are downstream of a community, that community's used water (run-off and treated wastewater) gets put back into river or stream and is delivered downstream to your community and becomes part of your drinking water supply.

Planned potable reuse is publicly acknowledged as an intentional project to recycle water for drinking water. It can be either direct or indirect. It commonly involves a more formal public process and public consultation program than is observed with de facto or unacknowledged reuse.

How potable reused water is delivered determines if it is called Indirect Potable Reuse or Direct Potable Reuse.

- Indirect potable reuse means the water is delivered to you indirectly. After it is purified, the reused water blends with other supplies and/or sits a while in some sort of storage, man-made or natural, before it gets delivered to a pipeline that leads to a water treatment plant or distribution system. That storage could be a groundwater basin or a surface water reservoir.
- Direct potable reuse means the reused water is put directly into pipelines that go to a water treatment plant or distribution system. Direct potable reuse may occur with or without "engineered storage" such as underground or above ground tanks.

Greywater uses the same waste as water reclamation with the exception of toilet water. Greywater can be used for irrigation, toilet flushing or other domestic uses.

Desalination is an energy-intensive process where salt and other minerals are removed from sea water to produce potable water for drinking and irrigation, typically through membrane filtration (reverse-osmosis), and steam-distillation.

Maximum water recovery - To determine maximum water recovery there are various techniques that have been developed by researchers; for maximum water reuse/reclamation/recovery strategies such as water pinch analysis. The techniques help a user to target the minimum freshwater consumption and wastewater target. It also helps in designing the network that achieves the target. This provides a benchmark to be used by users in improving their water systems.

Applications

Most of the uses of water reclamation are non potable uses such as: Washing Cars, flushing toilets, cooling water for power plants, concrete mixing, artificial lakes, irrigation for golf courses and public parks, and for hydraulic fracturing. Where applicable, systems run a dual piping system to keep the recycled water separate from the potable water.

New technologies for recycling allow the water to be used for fracking purposes and can save an estimated 4 - 7 million or more gallons per well.

Potable uses

Some water agencies reuse highly treated effluent from municipal wastewater or resource recovery plants as a reliable, drought proof source of drinking water. By using advanced purification processes, they produce water that meets all applicable drinking water standards. System reliability and frequent monitoring and testing are imperative to them meeting stringent controls.

The water needs of a community, water sources, public health regulations, costs, and the types of water infrastructure in place, such as distribution systems, man-made reservoirs, or natural groundwater basins, determine if and how reclaimed water can be part of the drinking water supply. Communities in El Paso, Texas and Orange County, California, for example, reuse water to replenish groundwater basins. Others, such as the Upper Occoquan Service Authority in Virginia, put it into surface water reservoirs. In these instances the reclaimed water is blended with other water supplies and/or sits in storage for a certain amount of time before it is drawn out and gets treated again at a water treatment or distribution system. In some Texas communities, the reused water is put directly into pipelines that go to a water treatment plant or distribution system. And in Singapore the reused water is bottled directly from an advanced water purification facility for educational and celebratory purposes. Though most of the reused water is used for high-tech industry in Singapore, a small amount is returned to reservoirs for drinking water.

Modern technologies such as reverse osmosis and ultraviolet disinfection are commonly used when reclaimed water will be mixed with the drinking water supply. An experiment by the University of New South Wales reportedly showed a reverse osmosis system removed ethinylestradiol and paracetamol from the wastewater, even at 1000 times the expected concentration.

Aboard the International Space Station, astronauts have been able to drink recycled urine due to the introduction of the ECLSS system. The system costs $250 million and has been working since May 2009. The system recycles wastewater and urine back into potable water used for drinking, food preparation, and oxygen generation. This cuts back on the need for resupplying the space station so often.

Indirect potable reuse

Some municipalities are using and others are investigating Indirect Potable Reuse (IPR) of reclaimed water. For example, reclaimed water may be pumped into (subsurface recharge) or percolated down to (surface recharge) groundwater aquifers, pumped out, treated again, and finally used as drinking water. This technique may also be referred to as groundwater recharging. This includes slow processes of further multiple purification steps via the layers of earth/sand (absorption) and microflora in the soil (biodegradation).

Direct potable reuse

In a Direct Potable Reuse (DPR) scheme, water is put directly into pipelines that go to a water treatment plant or distribution system. Direct potable reuse may occur with or without "engineered storage" such as underground or above ground tanks. Communities in Texas have implemented DPR projects, and the state of California is studying the feasibility of developing DPR regulations.

Unplanned potable reuse

Water reuse occurs in various ways throughout the world. It happens daily on rivers and other water bodies everywhere. If you live in a community downstream of another, chances are you are reusing its water and likewise communities downstream of you are most likely reusing your water. Unplanned

Indirect Potable Use has existed even before the introduction of reclaimed water. Many cities already use water from rivers that contain effluent discharged from upstream sewage treatment plants. There are many large towns on the River Thames upstream of London (Oxford, Reading, Swindon, Bracknell) that discharge their treated sewage ("non-potable water") into the river, which is used to supply London with water downstream.

DESIGN CONSIDERATIONS

Distribution

Nonpotable reclaimed water is often distributed with a dual piping network that keeps reclaimed water pipes completely separate from potable water pipes. In the United States and some other countries, nonpotable reclaimed water is distributed in lavender (light purple) pipes to distinguish it from potable water. The use of the color purple for pipes carrying recycled water was pioneered by the Irvine Ranch Water District in Irvine, California. In many cities using reclaimed water, it is now in such demand that consumers are only allowed to use it on assigned days. Some cities that previously offered unlimited reclaimed water at a flat rate are now beginning to charge citizens by the amount they use.

Reclamation processes

Wastewater must pass through numerous systems before being returned to the environment. Here is a partial listing from one particular plant system:

- Barscreens - Barscreens remove large solids that are sent into a grinder. All solids are then dumped into a sewer pipe at a Treatment Plant.
- Primary Settling Tanks - Readily settable and floatable solids are removed from the wastewater. These solids are skimmed from the top and bottom of the tanks and sent to the Treatment Plant where it'll be turned into fertilizer.
- Biological Treatment - The wastewater is cleaned through a biological treatment method that uses microorganisms, bacteria which digest the sludge and reduce the nutrient content. Air bubbles up to keep the organisms suspended and to supply oxygen to the aerobic bacteria so they can metabolize the food, convert it to energy, CO2, and water, and reproduce more microorganisms. This helps to remove ammonia also through nitrification.
- Secondary Settling Tanks - The force of the flow slows down as sewage enters these tanks, allowing the microorganisms to settle to the bottom.

As they settle, other small particles suspended in the water are picked up, leaving behind clear wastewater. Some of the microorganisms that settle to the bottom are returned to the system to be used again.

- Tertiary Treatment - Deep-bed, single-media, gravity sand filters receive water from the secondary basins and filter out the remaining solids. As this is the final process to remove solids, the water in these filters is almost completely clear.
- Chlorine Contact Tanks - Three chlorine contact tanks disinfect the water to decrease the risks associated with discharging wastewater containing human pathogens. This step protects the quality of the waters that receive the wastewater discharge.

One of two procedures are then followed according to the future disposal site:

1. Reclaimed Water Pump Station - The pump station distributes reclaimed water to users around the City. This may include golf courses, agricultural uses, cooling towers, or in land fills.
2. Water is passed through high level purification to be returned to the environment. Currently this means a reverse osmosis system.

TREATMENT IMPROVEMENTS

As world populations require both more clean water and better ways to dispose of wastewater, the demand for water reclamation will increase. Future success in water reuse will depend on whether this can be done without adverse effects on human health and the environment.

In the United States, reclaimed waste water is generally treated to secondary level when used for irrigation, but there are questions about the adequacy of that treatment. Some leading scientists in the main water society, AWWA, have long believed that secondary treatment is insufficient to protect people against pathogens, and recommend adding at least membrane filtration, reverse osmosis, ozonation, or other advanced treatments for irrigation water.

There have been recent advances in reverse osmosis in different countries, but have consistently produced very high quality water all the same. In Singapore, reclaimed water, also known as NEWater has become cleaner than the government issue tap water. Also, according to Bartels, the Bedok Demonstration Plant, which uses RO membranes, has successfully run for the past 3 years, producing high quality wastewater all the while. Seepage of nitrogen and phosphorus into ground and surface water is also becoming a serious problem, and will probably lead to at least tertiary treatment of reclaimed water to remove nutrients in the future. Even using secondary treatment, water quality can be improved. Water quality can also be improved as it passes through the subsurface mixing zone where surface water and groundwater combine. Testing for pathogens using Polymerase Chain Reaction (PCR) instead of older culturing techniques, and changing the discredited fecal coloform "indicator organism" standard would be improvements.

In a large study treatment plants showed that they could significantly reduce the numbers of parasites in effluent, just by making adjustments to the currently used process. But, even using the best of current technology, risk of spreading drug resistance in the environment through wastewater effluent, would remain.

Some scientists have suggested that there need to be basic changes in treatment, such as using bacteria to degrade waste based on nitrogen (urine) and not just carbonaceous (fecal) waste, saying that this would greatly improve effectiveness of treatment. Currently designed plants do not deal well with contaminants in solution (e.g. pharmaceuticals). "Dewatering" solids is a major problem. Some wastes could be disposed of without mixing them with water to begin with. In an interesting innovation, solids (sludge) could be removed before entering digesters and burned into a gas that could be used to run engines.

Emerging disinfection technologies include ultrasound, pulse arc electrohydraulic discharge, and bank filtration. Another issue is concern about weakened mandates for pretreatment of industrial wastes before they are made part of the municipal waste stream. Some also believe that hospitals should treat their own wastes. The safety of drinking reclaimed water which has been given advanced treatment and blended with other waters, remains controversial.

In recent years, as hydraulic fracturing of oil and gas formations has become more and more common place, new technologies for water recycling have emerged. One such technology uses a combination of ozone and electrocoagulation. This process removes organics, hydrocarbons, spent polymers, chemical additives used in the fracturing process, and heavy metals such as barium, iron, boron and more.

ALTERNATIVES

Seawater desalination

In urban areas where climate change has threatened long-term water security and reduced rainfall over catchment areas, using reclaimed water for indirect potable use may be superior to other water supply augmentation methods. One other commonly used option is seawater desalination. Recycling wastewater and desalinating seawater may have many of the same disadvantages, including high costs of water treatment, infrastructure construction, transportation, and waste disposal problems. Although the best option varies from region to region, desalination is often superior economically, as reclaimed water usually requires a dual piping network, often with additional storage tanks, when used for nonpotable use.

Greywater systems

A less elaborate alternative to reclaimed water is a greywater system. Greywater is wastewater that has been used in sinks, baths, showers, or washing machines, but does not contain sewage (see blackwater) and has not been treated at the same levels as recycled water. In a home system, treated or untreated greywater may be used to flush toilets or for irrigation. Some systems now exist which directly use greywater from a sink to flush a toilet.

Rainwater harvesting

Perhaps the simplest option is a rainwater harvesting system. Although there are concerns about the quality of rainwater in urban areas, due to air pollution and acid rain, many systems exist now to use untreated rainwater for nonpotable uses or treated rainwater for direct potable use. Urban design systems which incorporate rainwater harvesting and reduce runoff are known as Water Sensitive Urban Design (WSUD) in Australia, Low Impact Development (LID) in the United States and Sustainable urban drainage systems (SUDS) in the United Kingdom. There are also concerns about rainwater harvesting systems reducing the amount of run-off entering natural bodies of water.

Health aspects

Reclaimed water is highly engineered for safety and reliability so that the quality of reclaimed water is more predictable than many existing surface and groundwater sources. Reclaimed water is considered safe when appropriately used. Reclaimed water planned for use in recharging aquifers or augmenting surface water receives adequate and reliable treatment before mixing with naturally occurring water and undergoing natural restoration processes. Some of this water eventually becomes part of drinking water supplies.

A water quality study published in 2009 compared the water quality differences of reclaimed/recycled water, surface water, and groundwater. Results indicate that reclaimed water, surface water, and groundwater are more similar than dissimilar with regard to constituents. The researchers tested for

244 representative constituents typically found in water. When detected, most constituents were in the parts per billion and parts per trillion range. DEET (a bug repellant), and Caffeine were found in all water types and virtually in all samples. Triclosan (in anti-bacterial soap & toothpaste) was found in all water types, but detected in higher levels (parts per trillion) in reclaimed water than in surface or groundwater. Very few hormones/steroids were detected in samples, and when detected were at very low levels. Haloacetic acids (a disinfection by-product) were found in all types of samples, even groundwater. The largest difference between reclaimed water and the other waters appears to be that reclaimed water has been disinfected and thus has disinfection by-products (due to chlorine use).

A 2005 study titled "Irrigation of Parks, Playgrounds, and Schoolyards with Reclaimed Water" found that there had been no incidences of illness or disease from either microbial pathogens or chemicals, and the risks of using reclaimed water for irrigation are not measurably different from irrigation using potable water. Studies by the National Academies of Science, the Monterey Regional Water Pollution Control Agency, and others have found reclaimed water to be safe for agricultural use.

TESTING STANDARDS

Reclaimed water is not regulated by the Environmental Protection Agency (EPA), but the EPA has developed water reuse guidelines that were most recently updated in 2012. The EPA Guidelines for Water Reuse represents the international standard for best practices in water reuse. The document was developed under a Cooperative Research and Development Agreement between the U.S. Environmental Protection Agency (EPA), the U.S. Agency for International Development (USAID), and the global consultancy CDM Smith. The Guidelines provide a framework for states to develop regulations that incorporate the best practices and address local requirements.

Ongoing wastewater research sometimes raise concerns about pathogens in the water. Many pathogens cannot be detected by currently used tests. Recent literature also questions the validity of testing for "indicator organisms" instead of pathogens. Nor do present standards consider interactions of heavy metals and pharmaceuticals which may foster the development of drug resistant pathogens in waters derived from sewage.

To address these concerns about the source water, reclaimed water providers use multi-barrier treatment processes and constant monitoring to ensure that reclaimed water is safe and treated properly for the intended end use.

POTABLE USE

The main health risk for potable use of reclaimed water is the potential for pharmaceutical and other household chemicals or their derivatives (Environmental persistent pharmaceutical pollutants) to persist in this water. This would be of much less concern if the population were to keep their excrement out of the wastewater e.g. via the use of the Urine-diverting dry toilet or systems that treat blackwater separately from greywater.

Environmental aspects

There is debate about possible health and environmental effects. To address these concerns, A Risk Assessment Study of potential health risks of recycled water and comparisons to conventional Pharmaceuticals and Personal Care Product (PPCP) exposures was conducted by the WateReuse Research Foundation. For each of four scenarios in which people come into contact with recycled water used for irrigation - children on the playground, golfers, and landscape, and agricultural workers - the findings from the study indicate that it could take anywhere from a few years to millions of years of exposure to nonpotable recycled water to reach the same exposure to PPCPs that we get in a single day through routine activities. Using reclaimed water for non-potable uses saves potable water for drinking, since less potable water will be used for non-potable uses.

It sometimes contains higher levels of nutrients such as nitrogen, phosphorus and oxygen which may somewhat help fertilize garden and agricultural plants when used for irrigation.

The usage of water reclamation decreases the pollution sent to sensitive environments. It can also enhance wetlands, which benefits the wildlife depending on that eco-system. It also helps to stop the chances of drought as recycling of water reduces the use of fresh water supply from underground sources. For instance, The San Jose/Santa Clara Water Pollution Control Plant instituted a water recycling program to protect the San Francisco Bay area's natural salt water marshes.

Costs and evaluation

The cost of reclaimed water exceeds that of potable water in many regions of the world, where a fresh water supply is plentiful. However, reclaimed water is usually sold to citizens at a cheaper rate to encourage its use. As fresh water supplies become limited from distribution costs, increased population demands, or climate change reducing sources, the cost ratios will evolve also. The evaluation of reclaimed water needs to consider the entire water supply system, as it may bring important value of flexibility into the overall system

History

Storm and sanitary sewers were necessarily developed along with the growth of cities. By the 1840s the luxury of indoor plumbing, which mixes human waste with water and flushes it away, eliminated the need for cesspools. Odor was considered the big problem in waste disposal and to address it, sewage could be drained to a lagoon, or "settled" and the solids removed, to be disposed of separately. This process is now called "primary treatment" and the settled solids are called "sludge."

At the end of the 19th century, since primary treatment still left odor problems, it was discovered that bad odors could be prevented by introducing oxygen into the decomposing sewage. This was the beginning of the biological aerobic and anaerobic treatments which are fundamental to waste water processes.

By the 1920s, it became necessary to further control the pollution caused by the large quantities of human and industrial liquid wastes which were being piped into rivers and oceans, and modern treatment plants were being built in the US and other industrialized nations by the 1930s.

Designed to make water safe for fishing and recreation, the Clean Water Act of 1972 mandated elimination of the discharge of untreated waste from municipal and industrial sources, and the US federal government provided billions of dollars in grants for building sewage treatment plants around the country. Modern treatment plants, usually using oxidation and/or chlorination in addition to primary and secondary treatment, were required to meet certain standards.

Current treatment improves the quality of separated wastewater solids or sludge. The separated water is given further treatment considered adequate for non potable use by local agencies, and discharged into bodies of water, or reused as reclaimed water. In places like Florida, where it is necessary to avoid nutrient overload of sensitive receiving water, reuse of treated or reclaimed water can be more economically feasible than meeting the higher standards for surface water disposal mandated by the Clean Water Act.

7

CANAL BUILDING

Canals and navigations are human-made channels for water conveyance (supply), or to service water transport vehicles. In the vernacular, both are referred to as 'canals', and in most cases, the engineered works will have a series of dams and locks that create areas of low speed current flow. These areas are referred to as 'slack water levels', often just called 'levels'.

The main difference between them is that a navigation parallels a river and shares part of its waters, its drainage basin and leverages its resources by building dams and locks to increase and lengthen its long stretches of deep enough slack water levels while staying in its valley. In contrast, a canal cuts across a drainage divide atop a ridge (however humble), generally requiring an external water source above the highest elevation. Many canals have been built at elevations towering over valleys and others water ways crossing far below. Cities need a lot of fresh water and many canals with sources of water at a higher level can deliver water to a destination where there is a lack of water. The Roman Empire's Aqueducts were such water supply canals.

CANALS

Canals are man-made channels for water. There are two types of canal:

1. Aqueducts: water supply canals that are used for the conveyance and delivery of potable water for human consumption, municipal uses, and agriculture irrigation. Rills and Acequias are small versions.

2. Waterways: navigable transportation canals used for carrying ships and boats shipping goods and conveying people.

- Those connected to existing lakes, rivers, or oceans. Included are inter-basin canals, such as the Suez Canal, Erie Canal, and the Panama Canal.
- Those connected in a city network: such as the Canal Grande and others of Venice Italy; the gracht of Amsterdam, and the waterways of Bangkok Thailand.

TYPES OF ARTIFICIAL WATERWAYS

Canals are created in one of three ways, or a combination of the three, depending on available water and available path:

- A canal can be created where no stream presently exists. The body of the canal is either dug or the sides are created by piling dirt, stone, concrete, or other building materials. The water for the canal must be provided from an external source like other streams or reservoirs. Examples include canals that connect valleys over a higher body of land, like Canal du Midi and Canal de Briare.
- A stream can be canalized to make its navigable path more predictable and easier to maneuver. Canalization modifies the stream to more safely carry traffic by controlling the flow of the stream with dredging, damming, and modifying its path. Examples include Basse Saône, Canal de Mines de Fer de la Moselle, and Aisne River. Riparian zone restoration may be required.
- When a stream is too difficult to modify with canalization, a second stream can be created next to the existing stream. This is called a lateral canal. The existing stream usually acts as the water source and its banks provide a path for the new body. Examples include the Chesapeake and Ohio Canal, Canal latéral à la Loire, Garonne Lateral Canal, and Canal latéral à l'Aisne.

Smaller transportation canals can carry barges or narrowboats, while ship canals allow seagoing ships to travel to an inland port (e.g.: Manchester Ship Canal), or from one sea or ocean to another (e.g.: Caledonian Canal, Panama Canal).

Features

At their simplest, canals consist of a trench filled with water. Depending on the stratum the canal passes through, it may be necessary to line the cut with some form of watertight material such as clay or concrete. When this is done with clay this is known as puddling.

Canals need to be level, and while small irregularities in the lie of the land can be dealt with through cuttings and embankments, for larger deviations, other approaches have been adopted. The most common is the pound lock which consists of a chamber within which the water level can be raised or lowered connecting either two pieces of canal at a different level or the canal with a river or the sea. When there is a hill to be climbed, flights of many locks in short succession may be used.

Prior to the development of the pound lock in 984AD in China by Chhaio Wei-Yo and later in Europe in the 15th century, either flash locks consisting of a single gate were used or ramps, sometimes equipped with rollers, were used to change level. Flash locks were only practical where there was plenty of water available.

Locks use a lot of water, so builders have adopted other approaches. These include boat lifts, such as the Falkirk wheel, which use a caisson of water in which boats float while being moved between two levels; and inclined planes where a caisson is hauled up a steep railway.

To cross a stream or road, the solution is usually to bridge with an aqueduct. To cross a wide valley (where the journey delay caused by a flight of locks at either side would be unacceptable) the centre of the valley can be spanned by an aqueduct - a famous example in Wales is the Pontcysyllte aqueduct across the valley of the River Dee.

Another option for dealing with hills is to tunnel through them. An example of this approach is the Harecastle Tunnel on the Trent and Mersey Canal. Tunnels are only practical for smaller canals.

Some canals attempted to keep changes in level down to a minimum. These canals known as contour canals would take longer winding routes, along which the land was a uniform altitude. Other generally later canals took more direct routes requiring the use of various methods to deal with the change in level.

Canals have various features to tackle the problem of water supply. In some cases such as the Suez Canal the canal is simply open to the sea. Where the canal is not at sea level a number of approaches have been adopted. Taking water from existing rivers or springs was an option in some cases, sometimes supplemented by other methods to deal with seasonal variations in flow. Where such sources were unavailable, reservoirs, either separate from the canal, or built into its course, and back pumping was used to provide the required water. In other cases water pumped from mines was used to feed the canal. In certain cases extensive "feeder canals" were built to bring water from sources located far from the canal.

Where large amounts of goods are loaded or unloaded such as at the end of a canal a canal basin may be built. This would normally be a section of water wider than the general canal. In some cases the canal basins contain wharfs and cranes to assist with movement of goods.

When a section of the canal needs to be sealed off so it can be drained for maintenance stop planks are frequently used. These consist of planks of wood placed across the canal to form a dam. They are generally placed in pre existing grooves in the canal bank. On more modern canals, "guard locks" or gates were sometimes placed to allow a section of canal to be quickly closed off, either for maintenance, or to prevent a major loss of water due to a canal breach.

The oldest known canals were irrigation canals, built in Mesopotamia circa 4000 BC, in what is now modern day Iraq and Syria. The Indus Valley Civilization, in Pakistan and North India, (circa 2600 BC) had sophisticated irrigation and storage systems developed, including the reservoirs built at Girnar in 3000 BC. In Egypt, canals date back at least to the time of Pepi I Meryre (reigned 2332-2283 BC), who ordered a canal built to bypass the cataract on the Nile near Aswan.

In ancient China, large canals for river transport were established as far back as the Warring States (481-221 BC), the longest one of that period being the Hong Gou (Canal of the Wild Geese), which according to the ancient historian Sima Qian connected the old states of Song, Zhang, Chen, Cai, Cao, and Wei.

By far the longest canal was the Grand Canal of China, still the longest canal in the world today. It is 1,794 kilometres (1,115 mi) long and was built to carry the Emperor Yang Guang between Beijing and Hangzhou. The project began in 605 and was completed in 609, although much of the work combined older canals, the oldest section of the canal existing since at least 486 BC. Even in its narrowest urban sections it is rarely less than 30 metres (98 ft) wide. Greek engineers were the first to use canal locks, by which they regulated the water flow in the Ancient Suez Canal as early as the 3rd century BC.

Canals in the Middle Ages

In the Middle Ages, water transport was cheaper and faster than transport overland. This was because roads were unpaved and in poor condition and greater amounts could be transported by ship. The first artificial canal in Christian Europe was the Fossa Carolina built at the end of the 8th Century under personal supervision of Charlemagne. More lasting and of more economic impact were canals like the Naviglio Grande built between 1127 and 1257, the most important of the lombard “navigli”, Later, canals were built in the Netherlands and Flanders to drain the polders and assist the transportation of goods.

Canal building was revived in this age because of commercial expansion from the 12th century AD. River navigations were improved progressively by the use of single, or flash locks. Taking boats through these used large amounts of water leading to conflicts with watermill owners and to correct this, the pound or chamber lock first appeared, in 10th century AD in China and in Europe in 1373 in Vreeswijk, Netherlands. Another important development was the mitre gate which was probably introduced in Italy by Bertola da Novate in the sixteenth century. This allowed wider gates and also removed the height restriction of guillotine locks.

To break out of the limitations caused by river valleys, the first summit level canals were developed with the Grand Canal of China in 581-617 AD whilst in Europe the first, also using single locks, was the Stecknitz Canal in Germany in 1398. The first to use pound locks was the Briare Canal connecting the Loire and Seine (1642), followed by the more ambitious Canal du Midi (1683) connecting the Atlantic to the Mediterranean. This included a staircase of 8 locks at Béziers, a 157 metres (515 ft) tunnel and three major aqueducts.

Canal building progressed steadily in Germany in the 17th and 18th centuries with three great rivers, the Elbe, Oder and Weser being linked by canals. In post-Roman Britain, the first canal built appears to have been the Exeter Canal, which opened in 1563. The oldest canal built for industrial purposes in North America is Mother Brook in Dedham, MA. It was constructed in 1639 to provide water power for mills. In Russia, the Volga-Baltic Waterway, a nationwide canal system connecting the Baltic and Caspian seas via the Neva and Volga rivers, was opened in 1718.

INDUSTRIAL REVOLUTION

Canals were important for industrial development. The greatest stimulus to canal system building came from the Industrial Revolution with its need for cheap transport of unprecedented quantities of raw materials and manufactured items.

In Europe, particularly Britain and Ireland, and then in the young United States and the Canadian colonies, inland canals preceded the development of railroads during the earliest phase of the Industrial Revolution. The opening of the Sankey Canal in 1757, followed by the Bridgewater Canal in 1761, which halved the price of coal in Liverpool and Manchester, respectively, triggered a period of "canal mania" in Britain so that between 1760 and 1820 over one hundred canals were built.

The Blackstone Canal in Massachusetts and Rhode Island fulfilled a similar role in the early industrial revolution between 1828-48. The Blackstone Valley was a major contributor of the American Industrial Revolution where Samuel Slater built his first mill.

In addition to their transportation purposes, parts of the United States, particularly in the Northeast, had enough fast-flowing rivers that water power was the primary means of powering factories (usually textile mills) until after the American Civil War. For example, Lowell, Massachusetts, considered to be "The Cradle of the American Industrial Revolution," has 6 miles (9.7 km) of canals, built from around 1790 to 1850, that provided water power and a means of transportation for the city. The output of the system is estimated at 10,000 horsepower. Other cities with extensive power canal systems include Lawrence, Massachusetts, Holyoke, Massachusetts, Manchester, New Hampshire, and Augusta, Georgia.

THE 19TH CENTURY

Competition from the railway network from the 1830s, and in the 20th century the roads, made the smaller canals obsolete for most commercial transportation, and many of the British canals fell into decay. Only the Manchester Ship Canal and the Aire and Calder Canal bucked this trend. Yet in other countries canals grew in size as construction techniques improved. During the 19th century in the US, the length of canals grew from 100 miles (161 km) to over 4,000, with a complex network making the Great Lakes navigable, in conjunction with Canada, although some canals were later drained and used as railroad rights-of-way.

In the United States, navigable canals reached into isolated areas and brought them in touch with the world beyond. By 1825 the Erie Canal, 363 miles (584 km) long with 82 locks, opened up a connection from the populated Northeast to the Great Lakes. Settlers flooded into regions serviced by such canals, since access to markets was available. The Erie Canal (as well as other canals) was instrumental in lowering the differences in commodity prices between these various markets across America.

The canals caused price convergence between different regions because of their reduction in transportation costs, which allowed Americans to ship and buy goods from farther distances for much lower prices compared to before. Ohio built many miles of canal, Indiana had working canals for a few decades, and the Illinois and Michigan Canal connected the Great Lakes to the Mississippi River system until replaced by a channelized river waterway.

Three major canals with very different purposes were built in what is now Canada. The first Welland Canal, which opened in 1829 between Lake Ontario and Lake Erie, bypassing Niagara Falls and the Lachine Canal (1825) which allowed ships to skirt the nearly impassable rapids on the St. Lawrence River at Montreal were built for commerce.

The Rideau Canal, completed in 1832, connects Ottawa, on the Ottawa River to Kingston, Ontario on Lake Ontario. The Rideau Canal was built as a result of the War of 1812 to provide military transportation between the British colonies of Upper Canada and Lower Canada as an alternative to part of the St. Lawrence River which was susceptible to blockade by the United States.

In France, a steady linking of all the river systems-Rhine, Rhône, Saône and Seine-and the North Sea was boosted in 1879 by the establishment of the Freycinet gauge which specified the minimum size of locks so that canal traffic doubled in the first decades of the 20th century.

Many notable sea canals were completed in this period, starting with the Suez Canal (1869), and the Kiel Canal (1897), which carries tonnage many times that of most other canals, though the Panama Canal was not opened until 1914.

In the 19th century, a number of canals were built in Japan including the Biwako canal and the Tone canal. These canals were partially built with the help of engineers from the Netherlands and other countries.

Modern uses

Large scale ship canals such as the Panama Canal and Suez Canal continue to operate for cargo transportation; as do European barge canals. Due to globalization, they are becoming increasingly important, resulting in expansion projects such as the Panama Canal expansion project.

The narrow early industrial canals, however, have ceased to carry significant amounts of trade and many have been abandoned to navigation, but may still be used as a system for transportation of untreated water. In some cases railways have been built along the canal route, an example being the Croydon Canal.

A movement that began in Britain and France to use the early industrial canals for pleasure boats, such as hotel barges, has spurred rehabilitation of stretches of historic canals. In some cases abandoned canals such as the Kennet and Avon Canal have been restored and are now used by pleasure boaters. In Britain canalside housing has also proven popular in recent years.

The Seine-Nord Europe Canal is being developed into a major transportation waterway, linking France with Belgium, Germany and the Netherlands.

Canals have found another use in the 21st century, as easements for the installation of fibre optic telecommunications network cabling avoiding having them buried in roadways while facilitating access and reducing the hazard of being damaged from digging equipment.

Canals are still used to provide water for agriculture. An extensive canal system exists within the Imperial Valley in the Southern California desert to provide irrigation to agriculture within the area.

Canals are so deeply identified with Venice that many canal cities have been nicknamed "the Venice of…". The city is built on marshy islands, with wooden piles supporting the buildings, so that the land is man-made rather than the waterways. The islands have a long history of settlement; by the 12th century, Venice was a powerful city state.

Amsterdam was built in a similar way, with buildings on wooden piles. It became a city around 1300.Other cities with extensive canal networks include: Delft, Haarlem and Leiden in the Netherlands, Brugge in Flanders, Belgium, Birmingham in England - which has 35 miles of canals, to Venice's 26 miles -, Saint Petersburg in Russia, Hamburg in Germany, Berlin in Germany , Fort Lauderdale, Florida, and Cape Coral, Florida in the United States.

Liverpool Maritime Mercantile City is a UNESCO World Heritage Site near the centre of Liverpool, England, where a system of intertwining waterways and docks now being developed for mainly residential and leisure use.

Canal Estates (commonly known as bayous) are a form of subdivision popular in cities like Miami, Florida, Texas City, Texas and the Gold Coast, Queensland; the Gold Coast has over 700 km of residential canals. Wetlands are difficult areas upon which to build housing estates, so dredging part of the wetland down to a navigable channel provides fill to build up another part of the wetland above the flood level for houses. Land is built up in a finger pattern that provides a suburban street layout of waterfront housing blocks.

Inland canals have often had boats specifically built for them. An example of this is the British narrowboat, which is up to 72 feet (21.95 m) long and 7 feet (2.13 m) wide and was primarily built for British Midland canals. In this case the limiting factor was the size of the locks. This is also the limiting factor on the

Panama canal where Panamax ships are limited to a length of 294.1 m (965 ft) and a width of 32.3 m (106 ft).

For the lockless Suez Canal the limiting factor for Suezmax ships is generally draft, which is limited to 16 m (52.5 ft). At the other end of the scale, tub-boat canals such as the Bude Canal were limited to boats of under 10 tons for much of their length due to the capacity of their inclined planes or boat lifts. Most canals have a limit on height imposed either by bridges or tunnels.

- Dune ridges: In areas of large-scale dune activity, certain types of dunes result in sand ridges.
- Moraines and eskers: Glacial activity may leave ridges in the form of moraines and eskers. An arête is a thin ridge of rock that is formed by glacial erosion.
- Volcanic subglacial ridges: Many subglacial volcanoes create ridgelike formations when lava erupts through a thick glacier or ice sheet.

TECTONIC UPLIFT

Tectonic uplift is a geological process most often caused by plate tectonics which increases elevation. The opposite of uplift is subsidence, which results in a decrease in elevation. Uplift may be orogenic or isostatic. Orogenic uplift is the result of tectonic-plate collisions and results in mountain ranges or a more modest uplift over a large region. The Himalayas were (and are still being) formed by the collision of two continental plates, the Indian and Eurasian Plates. This ongoing collision produced the Tibetan Plateau as well as the Himalayas and associated ranges.

Ozark Plateau is a broad uplifted area which resulted from the Ouachita Orogeny to the south during the Permian Period. Another related uplift is the Llano Uplift in Texas, a geographical location named after its uplift features. The Colorado Plateau with its spectacular scenic canyons, the Grand Canyon, is also the result of broad tectonic uplift followed by river erosion.

ISOSTATIC UPLIFT

Isostatic uplift includes the gradual uplift following rapid erosional removal of material from a mountain range. The land rises as a result of the removal of the weight. Another example of isostatic uplift is post-glacial rebound following the melting of continental glaciers and ice sheets. The Hudson Bay region of Canada and the Great Lakes of Canada and the United States are currently undergoing gradual rebound as a result of the melting of the ice sheets 10,000 years ago.

Coral islands

In a few cases, tectonic uplift can be seen in the cases of coral islands. This is evidenced by the presence of various oceanic islands composed entirely of coral, which otherwise appear to be high islands (i.e., islands of volcanic origin). Examples of such islands are found in the Pacific, notably the three phosphate islets, Nauru, Makatea, and Banaba as well as Maré and Lifou in New-Caledonia, Fatu Huku in the Marquesas Islands and Henderson Island in the Pitcairn Islands. The uplift of these islands is the result of the movement of oceanic tectonic plates. Sunken islands or guyots with their coral reefs are the result of crustal subsidence as the oceanic plate carries the islands to deeper or lower oceanic crust areas.

Waterlogging

It refers to the saturation of soil with water. Soil may be regarded as waterlogged when the water table of the groundwater is too high to conveniently permit an anticipated activity, like agriculture.

In agriculture, various crops need air (specifically, oxygen) to a greater or lesser depth in the soil. Waterlogging of the soil stops air getting in. How near the water table must be to the surface for the ground to be classed as waterlogged, varies with the purpose in view. A crop's demand for freedom from waterlogging may vary between seasons of the year, as with the growing of rice (Oryza sativa).

In irrigated agricultural land, waterlogging is often accompanied by soil salinity as waterlogged soils prevent leaching of the salts imported by the irrigation water.

From a gardening point of view, waterlogging is the process whereby the soil blocks off all water and is so hard it stops air getting in and it stops oxygen from getting in.

Soil salinity control

Soil salinity control relates to controlling the problem of soil salinity and reclaiming salinized agricultural land. The aim of soil salinity control is to prevent soil degradation by salinization and reclaim already salty (saline) soils. Soil reclamation is also called soil improvement, rehabilitation, remediation, recuperation, or amelioration.

The primary man-made cause of salinization is irrigation. River water or groundwater used in irrigation contains salts, which remain behind in the soil after the water has evaporated.

The primary method of controlling soil salinity is to permit 10-20% of the irrigation water to leach the soil, be drained and discharged through an appropriate drainage system. The salt concentration of the drainage water is normally 5 to 10 times higher than that of the irrigation water, thus salt export matches salt import and it will not accumulate.

Salty (saline) soils are soils that have a high salt content. The predominant salt is normally sodium chloride (NaCl, "table salt"). Saline soils are therefore also sodic soils but there may be sodic soils that are not saline, but alkaline.

Salty soils are a common feature and an environmental problem in irrigated lands in arid and semi-arid regions. They have poor or little crop production. The problems are often associated with high water tables, caused by a lack of natural subsurface drainage to the underground. Poor subsurface drainage may be caused by insufficient transport capacity of the aquifer or because water cannot exit the aquifer for instance, if it is situated in a topographical depression.

Worldwide, the major factor in the development of saline soils is a lack of precipitation. Most naturally saline soils are found in (semi)arid regions and climates of the globe.

Primary cause

The primary cause of man-made salinization is the salt brought in with irrigation water. All irrigation water derived from rivers or groundwater, however 'sweet', contains salts that remain behind in the soil after the water has evaporated, which may create the environmental issue.

For example, assuming irrigation water with a low salt concentration of 0.3 g/l (equal to 0.3 kg/m^3 corresponding to an electric conductivity of about 0.5 dS/m) and a modest annual supply of irrigation water of 10,000 m^3/ha (almost 3 mm/day) brings 3,000 kg salt/ha each year. In the absence of sufficient natural drainage (as in waterlogged soils) and without a proper leaching and drainage program to remove salts, this would lead to a high soil salinity and reduced crop yields in the long run.

Much of the water used in irrigation has a higher salt content than in this example, which is compounded by that fact that many irrigation projects use a far greater annual supply of water. Sugar cane, for example, needs about 20000 m3/ha of water per year. As a result, irrigated areas often receive more than 3,000 kg/ha of salt per year and some receive as much as 10,000 kg/ha/year.

Secondary cause

The secondary cause of salinization is waterlogging in irrigated land.

Irrigation causes changes to the natural water balance of irrigated lands. Large quantities of water in irrigation projects are not consumed by plants and must go somewhere. In irrigation projects it is impossible to achieve 100% irrigation efficiency where all the irrigation water is consumed by the plants. The maximum attainable irrigation efficiency is about 70% but usually it is less than 60%. This means that minimum 30%, but usually more than 40% of the irrigation water is not evaporated and it must go somewhere.

Most of the water lost this way is stored underground which can change the original hydrology of local aquifers considerably. Many aquifers cannot absorb and transport these quantities of water and, the water table rises leading to water logging.

Waterlogging causes three problems:

- the shallow water table and lack of oxygenation of the root zone reduces the yield of most crops
- it leads to an accumulation of salts brought in with the irrigation water as their removal through the aquifer is blocked
- with the upward seepage of groundwater more salts are brought into the soil and the salination is aggravated

Aquifer conditions in irrigated land and the groundwater flow have an important role in soil salinization, as illustrated here :

- Illustration of the influence of aquifer conditions on soil salinization in irrigated land
- Soil salinization in the lower parts of undulating land with a good aquifer
- Soil salinization in the unirrigated parts of flat land with a good aquifer
- Soil salinization in irrigated flat land without an aquifer

Salt affected area

Normally, the salinization of agricultural land affects a considerable area of irrigation projects, on the order of 20 to 30%. When the agriculture in such a fraction of the land is abandoned, a new salt and water balance is attained, a new equilibrium is reached, and the situation becomes stable.

In India alone, thousands of square kilometres have been severely salinized. China and Pakistan do not lag much behind (perhaps China has even more salt affected land than India). A regional distribution of the 3,230,000 km^2 of saline land world wide is shown in the following table derived from the FAO/UNESCO Soil Map of the World.

Spatial variation

Although the principles of the processes of salinization are fairly easy to understand, it is more difficult to explain why certain parts of the land suffer from the problems and other parts do not, or to predict accurately which part of the land will fall victim. The main reason for this is the variation of natural conditions in time and space, the usually uneven distribution of the irrigation water, and the seasonal or yearly changes of agricultural practices.

Only in lands with undulating topography the explanation and prediction is pretty simple: the depressional areas will degrade the most.

The preparation of salt and water balances for distinguishable sub-areas in the irrigation project, or the use of agro-hydro-salinity models, can be helpful in explaining or predicting the extent and severity of the problems.

DIAGNOSIS

Measurement

Soil salinity is measured as the salt concentration of the soil solution in tems of g/l or electric conductivity (EC) in dS/m. The relation between these two units is about 5/3 : y g/l => 5y/3 dS/m. Sea water may have a salt concentration of 30 g/l (3%) and an EC of 50 dS/m. The standard for the determination of soil salinity is from an extract of a saturated paste of the soil, and the EC is then written as ECe. The extract is obtained by centrifugation. The salinity can more easily be measured, without centrifugation, in a 2:1 or 5:1 water:soil mixture (in terms of g water per g dry soil) than from a saturated paste. The relation between ECe and EC2:1 is about 4, hence : ECe = 4 EC1:2.

Classification

Soils are considered saline when the ECe > 4. When 4 < ECe < 8, the soil is called slightly saline, when 8 < ECe < 16 it is called (moderately) saline, and when ECe > 16 severely saline.

CROP TOLERANCE

Sensitive crops loose their vigor already in slightly saline soils, most crops are negatively affected by (moderately) saline soils, and only salinity resistant crops thrive in severely saline soils. The University of Wyoming and the Government of Alberta report data on the salt tolerance of plants.

Principles of salinity control

Drainage is the primary method of controlling soil salinity. The system should permit a small fraction of the irrigation water (about 10 to 20 percent, the drainage or leaching fraction) to be drained and discharged out of the irrigation project.

In irrigated areas where salinity is stable, the salt concentration of the drainage water is normally 5 to 10 times higher than that of the irrigation water. Salt export matches salt import and salt will not accumulate.

When reclaiming already salinized soils, the salt concentration of the drainage water will initially be much higher than that of the irrigation water (for example 50 times higher). Salt export will greatly exceed salt import, so that with the same drainage fraction a rapid desalinization occurs. After one or two years, the soil salinity is decreased so much, that the salinity of the drainage water has come down to a normal value and a new, favorable, equilibrium is reached.

In regions with pronounced dry and wet seasons, the drainage system may be operated in the wet season only, and closed during the dry season. This practice of checked or controlled drainage saves irrigation water.

The discharge of salty drainage water may pose environmental problems to downstream areas. The environmental hazards must be considered very carefully and, if necessary mitigating measures must be taken. If possible, the drainage must be limited to wet seasons only, when the salty effluent inflicts the least harm.

Drainage systems

The drainage system designed to evacuate salty water also lowers the water table. To reduce the cost of the system, the lowering must be reduced to a minimum. The highest permissible level of the water table (or the shallowest permissible depth) depends on the irrigation and agricultural practices and kind of crops.

In many cases a seasonal average water table depth of 0.6 to 0.8 m is deep enough. This means that the water table may occasionally be less than 0.6 m (say 0.2 m just after an irrigation or a rain storm). This automatically implies that, in other occasions, the water table will be deeper than 0.8 m (say 1.2 m). The fluctuation of the water table helps in the breathing function of the soil while the expulsion of carbon dioxide (CO_2) produced by the plant roots and the inhalation of fresh oxygen (O_2) is promoted.

The establishing of a not too deep water table offers the additional advantage that excessive field irrigation is discouraged, as the crop yield would be negatively affected by the resulting elevated water table, and irrigation water may be saved.

The statements made above on the optimum depth of the watertable are very general, because in some instances the required water table may be still shallower than indicated (for example in rice paddies), while in other instances it must be considerably deeper (for example in some orchards). The establishment of the optimum depth of the water table is in the realm of agricultural drainage criteria.

Soil leaching

Water balance factors in the soil

The unsaturated zone or vadose zone of the soil below the soil surface and the watertable is subject to four main hydrological inflow and outflow factors :

- Infiltration of rain and irrigation water (Irr) into the soil through the soil surface (Inf) :
- Inf = Rain + Irr
- Evaporation of soil water through plants and directly into the air through the soil surface (Evap)
- Percolation of water from the unsaturated zone soil into the groundwater through the watertable (Perc)
- Capillary rise of groundwater moving by capillary suction forces into the unsaturated zone(Cap)

In steady state (i.e. the amount of water stored in the unsaturated zone does not change in the long run) the water balance of the unsaturated zone reads: Inflow = Outflow, thus:

- Inf + Cap = Evap + Perc or :
- Irr + Rain + Cap = Evap + Perc and the salt balance is
- Irr.Ci + Cap.Cc = Evap.Fc.Ce + Perc.Cp + Ss

where Ci is the salt concentration of the irrigation water, Cc is the salt concentration of the capillary rise, equal to the salt concentration of the upper part of the groundwater body, Fc is the fraction of the total evaporation transpired by plants, Ce is the salt concentration of the water taken up by the plant roots, Cp is the salt concentration of the percolation water, and Ss is the increase of salt storage in the unsaturated soil. This assumes that the rainfall contains no salts. Only along the coast this may not be true. Further it is assumed that no runoff or surface drainage occurs.

The amount of salts removed by plants (Evap.Fc.Ce) is usually negligibly small: Evap.Fc.Ce = 0

The salt concentration Cp can be taken as a part of the salt concentration of the soil in the unsaturated zone (Cu) giving: Cp=Le.Cu, where Le is the leaching efficiency. The leaching efficiency is often in the order of 0.7 to 0.8, but in poorly structured, heavy clay soils it may be less. In the Leziria Grande polder in the delta of the Tagus river in Portugal it was found that the leaching efficiency was only 0.15.

Assuming that one wishes to avoid the soil salinity to increase and maintain the soil salinity Cu at a desired level Cd we have:

Ss = 0, Cu = Cd and Cp = Le.Cd. Hence the salt balance can be simplified to:

- Perc.Le.Cd = Irr.Ci + Cap.Cc

Setting the amount percolation water required to fulfill this salt balance equal to Lr (the leaching requirement) it is found that:

- Lr = (Irr.Ci + Cap.Cc) / Le.Cd .

Substituting herein Irr = Evap + Perc ? Rain ? Cap and re-arranging gives :

- Lr = [(Evap?Rain).Ci + Cap(Cc?Ci)] / (Le.Cd ? Ci)

With this the irrigation and drainage requirements for salinity control can can be computed too.

In irrigation projects in (semi)arid zones and climates it is important to check the leaching requirement, whereby the field irrigation efficiency (indicating the fraction of irrigation water percolating to the underground) is to be taken into account.

The desired soil salinity level Cd depends on the crop tolerance to salt. The University of Wyoming, USA, and the Government of Alberta, Canada, report crop tolerance data.

Strip cropping: an alternative

In irrigated lands with scarce water resources suffering from drainage (high water table) and soil salinity problems, strip cropping is sometimes practiced with strips of land where every other strip is irrigated while the strips in between are left permanently fallow.

Owing to the water application in the irrigated strips they have a higher watertable which induces flow of groundwater to the unirrigated strips. This flow functions as subsurface drainage for the irrigated strips, whereby the water table is maintained at a not-too-shallow depth, leaching of the soil is possible, and the soil salinity can be controlled at an acceptably low level.

In the unirrigated (sacrificial) strips the soil is dry and the groundwater comes up by capillary rise and evaporates leaving the salts behind, so that here the soil salinizes. Nevertheless, they can have some use for livestock, sowing salinity resistant grasses or weeds. Moreover, useful salt resistant trees can be planted like Casuarina, Eucalyptus or Atriplex, keeping in mind that the trees have deep rooting systems and the salinity of the wet subsoil is less than of the topsoil. In these ways wind erosion can be controlled. The unirrigated strips can also be used for salt harvesting.

The majority of the computer models available for water and solute transport in the soil (e.g. SWAP, DrainMod-S, UnSatChem), are based on Richard's differential equation for the movement of water in unsaturated soil in combination with a differential salinity dispersion equation.

The models require input of soil characteristics like the relation between unsaturated soil moisture content, water tension, hydraulic conductivity and dispersivity. These relations vary to a great extent from place to place and are not easy to measure. The models use short time steps and need at least a daily data base of hydrological phenomena. Altogether this makes model application to a fairly large project the job of a team of specialists with ample facilities.

Simpler models, like SaltMod, based on seasonal water and soil balances and an empirical capillary rise function, are also available. They are useful for long-term salinity predictions in relation to irrigation and drainage practices.Spatial variations owing to variations in topography can be simulated and predicted using salinity cum groundwater models, like SahysMod.

WATER RESOURCES

Water resources are sources of water that are useful or potentially useful to humans. Uses of water include agricultural, industrial, household, recreational and environmental activities. Virtually all of these human uses require fresh water.

97% of water on the Earth is salt water, and only 3% is fresh water of which slightly over two thirds is frozen in glaciers and polar ice caps. The remaining unfrozen freshwater is mainly found as groundwater, with only a small fraction present above ground or in the air.

Fresh water is a renewable resource, yet the world's supply of clean, fresh water is steadily decreasing. Water demand already exceeds supply in many parts of the world and as the world population continues to rise, so too does the water demand. Awareness of the global importance of preserving water for ecosystem services has only recently emerged as, during the 20th century, more than half the world's wetlands have been lost along with their valuable environmental services. Biodiversity-rich freshwater ecosystems are currently declining faster than marine or land ecosystems. The framework for allocating water resources to water users (where such a framework exists) is known as water rights.

SOURCES OF FRESH WATER

Surface water is water in a river, lake or fresh water wetland. Surface water is naturally replenished by precipitation and naturally lost through discharge to the oceans, evaporation, evapotranspiration and sub-surface seepage.

Although the only natural input to any surface water system is precipitation within its watershed, the total quantity of water in that system at any given time is also dependent on many other factors. These factors include storage capacity in lakes, wetlands and artificial reservoirs, the permeability of the soil beneath these storage bodies, the runoff characteristics of the land in the watershed, the timing of the precipitation and local evaporation rates. All of these factors also affect the proportions of water lost.

Human activities can have a large and sometimes devastating impact on these factors. Humans often increase storage capacity by constructing reservoirs and decrease it by draining wetlands. Humans often increase runoff quantities and velocities by paving areas and channelizing stream flow.

The total quantity of water available at any given time is an important consideration. Some human water users have an intermittent need for water. For example, many farms require large quantities of water in the spring, and no water at all in the winter. To supply such a farm with water, a surface water system may require a large storage capacity to collect water throughout the year and release it in a short period of time. Other users have a continuous need for water, such as a power plant that requires water for cooling. To supply such a power plant with water, a surface water system only needs enough storage capacity to fill in when average stream flow is below the power plant's need. Nevertheless, over the long term the average rate of precipitation within a watershed is the upper bound for average consumption of natural surface water from that watershed.

Natural surface water can be augmented by importing surface water from another watershed through a canal or pipeline. It can also be artificially augmented from any of the other sources listed here, however in practice the quantities are negligible. Humans can also cause surface water to be "lost" (i.e. become unusable) through pollution.

Brazil is the country estimated to have the largest supply of fresh water in the world, followed by Russia and Canada.

UNDER RIVER FLOW

Throughout the course of the river, the total volume of water transported downstream will often be a combination of the visible free water flow together with a substantial contribution flowing through sub-surface rocks and gravels that underlie the river and its floodplain called the hyporheic zone. For many rivers in large valleys, this unseen component of flow may greatly exceed the visible flow.

The hyporheic zone often forms a dynamic interface between surface water and true ground-water receiving water from the ground water when aquifers are fully charged and contributing water to ground-water when ground waters are depleted. This is especially significant in karst areas where pot-holes and underground rivers are common.

Ground water

Sub-surface water, or groundwater, is fresh water located in the pore space of soil and rocks. It is also water that is flowing within aquifers below the water table. Sometimes it is useful to make a distinction between sub-surface water that is closely associated with surface water and deep sub-surface water in an aquifer (sometimes called "fossil water").

Sub-surface water can be thought of in the same terms as surface water: inputs, outputs and storage. The critical difference is that due to its slow rate of turnover, sub-surface water storage is generally much larger compared to inputs than it is for surface water. This difference makes it easy for humans to use sub-surface water unsustainably for a long time without severe consequences. Nevertheless, over the long term the average rate of seepage above a sub-surface water source is the upper bound for average consumption of water from that source.

The natural input to sub-surface water is seepage from surface water. The natural outputs from sub-surface water are springs and seepage to the oceans. If the surface water source is also subject to substantial evaporation, a sub-surface water source may become saline. This situation can occur naturally under endorheic bodies of water, or artificially under irrigated farmland.

In coastal areas, human use of a sub-surface water source may cause the direction of seepage to ocean to reverse which can also cause soil salinization. Humans can also cause sub-surface water to be "lost" (i.e. become unusable) through pollution. Humans can increase the input to a sub-surface water source by building reservoirs or detention ponds.

Desalination

Desalination is an artificial process by which saline water (generally sea water) is converted to fresh water. The most common desalination processes are distillation and reverse osmosis. Desalination is currently expensive compared to most alternative sources of water, and only a very small fraction of total human use is satisfied by desalination. It is only economically practical for high-valued uses (such as household and industrial uses) in arid areas. The most extensive use is in the Persian Gulf.

Frozen water

Several schemes have been proposed to make use of icebergs as a water source, however to date this has only been done for novelty purposes. Glacier runoff is considered to be surface water.

The Himalayas, which are often called "The Roof of the World", contain some of the most extensive and rough high altitude areas on Earth as well as the greatest area of glaciers and permafrost outside of the poles. Ten of Asia's largest rivers flow from there, and more than a billion people's livelihoods depend on them. To complicate matters, temperatures are rising more rapidly here than the global average.

In Nepal the temperature has risen with 0.6 degree over the last decade, whereas the global warming has been around 0.7 over the last hundred years. Uses of fresh water can be categorized as consumptive and non-consumptive (sometimes called "renewable"). A use of water is consumptive if that water is not immediately available for another use. Losses to sub-surface seepage and evaporation are considered consumptive, as is water incorporated into a product (such as farm produce). Water that can be treated and returned as surface water, such as sewage, is generally considered non-consumptive if that water can be put to additional use.

Agricultural

It is estimated that 69% of worldwide water use is for irrigation, with 15-35% of irrigation withdrawals being unsustainable. It takes around 3,000 litres of water, converted from liquid to vapour, to produce enough food to satisfy one person's daily dietary need. This is a considerable amount, when compared to that required for drinking, which is between two and five litres. To produce food for the 6.5 billion or so people who inhabit the planet today requires the water that would fill a canal ten metres deep, 100 metres wide and 7.1 million kilometres long - that's enough to circle the globe 180 times.

Increasing water scarcity

Fifty years ago, the common perception was that water was an infinite resource. At this time, there were fewer than half the current number of people on the planet. People were not as wealthy as today, consumed fewer calories and ate less meat, so less water was needed to produce their food. They required a third of the volume of water we presently take from rivers. Today, the competition for water resources is much more intense.

This is because there are now nearly seven billion people on the planet, their consumption of water-thirsty meat and vegetables is rising, and there is increasing competition for water from industry, urbanisation and biofuel crops. In future, even more water will be needed to produce food because the Earth's population is forecast to rise to 9 billion by 2050. An additional 2.5 or 3 billion people, choosing to eat fewer cereals and more meat and vegetables could add an additional five million kilometres to the virtual canal mentioned above.

An assessment of water management in agriculture was conducted in 2007 by the International Water Management Institute in Sri Lanka to see if the world had sufficient water to provide food for its growing population. It assessed the current availability of water for agriculture on a global scale and mapped out locations suffering from water scarcity.

It found that a fifth of the world's people, more than 1.2 billion, live in areas of physical water scarcity, where there is not enough water to meet all demands. A further 1.6 billion people live in areas experiencing economic water scarcity, where the lack of investment in water or insufficient human capacity make it impossible for authorities to satisfy the demand for water.

The report found that it would be possible to produce the food required in future, but that continuation of today's food production and environmental trends would lead to crises in many parts of the world. To avoid a global water crisis, farmers will have to strive to increase productivity to meet growing demands for food, while industry and cities find ways to use water more efficiently.

In some areas of the world irrigation is necessary to grow any crop at all, in other areas it permits more profitable crops to be grown or enhances crop yield. Various irrigation methods involve different trade-offs between crop yield, water consumption and capital cost of equipment and structures. Irrigation methods such as furrow and overhead sprinkler irrigation are usually less expensive but are also typically less efficient, because much of the water evaporates, runs off or drains below the root zone.

Other irrigation methods considered to be more efficient include drip or trickle irrigation, surge irrigation, and some types of sprinkler systems where the sprinklers are operated near ground level. These types of systems, while more expensive, usually offer greater potential to minimize runoff, drainage and evaporation. Any system that is improperly managed can be wasteful, all methods have the potential for high efficiencies under suitable conditions, appropriate irrigation timing and management. Some issues that are often insufficiently considered are salinization of sub-surface water and contaminant accumulation leading to water quality declines.

As global populations grow, and as demand for food increases in a world with a fixed water supply, there are efforts under way to learn how to produce more food with less water, through improvements in irrigation methods and technologies, agricultural water management, crop types, and water monitoring. Aquaculture is a small but growing agricultural use of water. Freshwater commercial fisheries may also be considered as agricultural uses of water, but have generally been assigned a lower priority than irrigation (see Aral Sea and Pyramid Lake).

INDUSTRIAL

It is estimated that 22% of worldwide water use is industrial. Major industrial users include power plants, which use water for cooling or as a power source (i.e. hydroelectric plants), ore and oil refineries, which use water in chemical processes, and manufacturing plants, which use water as a solvent.

The portion of industrial water usage that is consumptive varies widely, but as a whole is lower than agricultural use. Water is used in power generation. Hydroelectricity is electricity obtained from hydropower. Hydroelectric power comes from water driving a water turbine connected to a generator. Hydroelectricity is a low-cost, non-polluting, renewable energy source. The energy is supplied by the sun. Heat from the sun evaporates water, which condenses as rain in higher altitudes, from where it flows down.

Pressurized water is used in water blasting and water jet cutters. Also, very high pressure water guns are used for precise cutting. It works very well, is relatively safe, and is not harmful to the environment. It is also used in the cooling of machinery to prevent over-heating, or prevent saw blades from over-heating.

Water is also used in many industrial processes and machines, such as the steam turbine and heat exchanger, in addition to its use as a chemical solvent. Discharge of untreated water from industrial uses is pollution. Pollution includes discharged solutes (chemical pollution) and discharged coolant water (thermal pollution). Industry requires pure water for many applications and utilizes a variety of purification techniques both in water supply and discharge.

HOUSEHOLD

It is estimated that 8% of worldwide water use is for household purposes. These include drinking water, bathing, cooking, sanitation, and gardening. Basic household water requirements have been estimated by Peter Gleick at around 50 liters per person per day, excluding water for gardens. Drinking water is water that is of sufficiently high quality so that it can be consumed or used without risk of immediate or long term harm.

Such water is commonly called potable water. In most developed countries, the water supplied to households, commerce and industry is all of drinking water standard even though only a very small proportion is actually consumed or used in food preparation.

Recreation

Recreational water use is usually a very small but growing percentage of total water use. Recreational water use is mostly tied to reservoirs. If a reservoir is kept fuller than it would otherwise be for recreation, then the water retained could be categorized as recreational usage. Release of water from a few reservoirs is also timed to enhance whitewater boating, which also could be considered a recreational usage. Other examples are anglers, water skiers, nature enthusiasts and swimmers.

TYPES OF ARTIFICIAL WATERWAYS

A navigation is a series of channels that run roughly parallel to the valley and stream bed of an unimproved river. A navigation always shares the drainage basin of the river. A vessel uses the calm parts of the river itself as well as improvements, traversing the same changes in height.

A true canal is a channel that cuts across a drainage divide, making a navigable channel connecting two different drainage basins.

Most commercially important canals of the first half of the 19th century were a little of each, using rivers in long stretches, and divide crossing canals in others. This is true for many canals still in use.

Structures used in artificial waterways

Both navigations and canals use engineered structures to improve navigation:

weirs and dams to raise river water levels to usable depths;

looping descents to create a longer and gentler channel around a stretch of rapids or falls;

locks to allow ships and barges to ascend/descend.

Since they cut across drainage divides, canals are more difficult to construct and often need additional improvements, like viaducts and aqueducts to bridge waters over streams and roads, and ways to keep water in the channel.

Types of canals

There are two broad types of canal:

Waterways: canals and navigations used for carrying vessels transporting goods and people. These can be subdivided into two kinds:

Those connecting existing lakes, rivers, other canals or seas and oceans.

Those connected in a city network: such as the Canal Grande and others of Venice Italy; the gracht of Amsterdam, and the waterways of Bangkok.

Aqueducts: water supply canals that are used for the conveyance and delivery of potable water for human consumption, municipal uses, hydro power canals and agricultureirrigation.

Loading Anthracite on the Lehigh Canal to feed the early United States industries in the pioneer-era.

1. Design High Water Level (HWL) 2. Low water channel 3. Flood channel 4. Riverside slope 5. Riverside banquette 6. Levee crown 7. Landside slope 8.

Landside banquette 9. Berm 10. Low water revetment 11. Riverside land 12. Levee 13. Protected lowland 14. River zone

IMPORTANCE

Historically canals were of immense importance to commerce and the development, growth and vitality of a civilization. In 1855 the Lehigh Canal carried over 1.2 million tons of clean burning anthracite coal; by the 1930s the company which built and operated it over a century pulled the plug. The few canals still in operation in our modern age are a fraction of the numbers that once fueled and enabled economic growth, indeed were practically a prerequisite to further urbanization and industrialization—for the movement of bulk raw materials such as coal and ores are difficult and marginally affordable without water transport. Such raw materials fueled the industrial developments and new metallurgy resulting of the spiral of increasing mechanization during 17th–20th century, leading to new research disciplines, new industries and economies of scale, raising the standard of living for any industrialized society.

The surviving canals, including most ship canals, today primarily service mostly bulk cargo and large ship transportation industries, whereas the once critical smaller inland waterways conceived and engineered as boat and barge canals have largely been supplanted and filled in, abandoned and left to deteriorate, or kept in service and staffed by state employees, where dams and locks are maintained for flood control or pleasure boating. Their replacement was gradual, beginning first in the United States in the mid-1850s where canal shipping was first augmented by, then began being replaced by using much faster, less geographically constrained & limited, and generally cheaper to maintain railways.

By the early 1880s, canals which had little ability to economically compete with rail transport, were off the map. In the next couple of decades, coal was increasingly diminished as the heating fuel of choice by oil, and growth of coal shipments levelled off. Later, after World War I when motor-trucks came into their own, the last small U.S. barge canals saw a steady decline in cargo ton-miles alongside many railways, the flexibility and steep slope climbing capability of lorries taking over cargo hauling increasingly as road networks were improved, and which also had the freedom to make deliveries well away from rail lined road beds or ditches in the dirt which couldn't operate in the winter.

Construction

Canals are built in one of three ways, or a combination of the three, depending on available water and available path:

Human made streams

A canal can be created where no stream presently exists. Either the body of the canal is dug or the sides of the canal are created by making dykes or levees by piling dirt, stone, concrete or other building materials. The finished shape of the canal as seen in cross section is known as the canal prism. The water for the canal must be provided from an external source, like streams or reservoirs. Where the new waterway must change elevation engineering works like locks, lifts or elevators are constructed to raise and lower vessels. Examples include canals that connect valleys over a higher body of land, like Canal du Midi, Canal de Briare and the Panama Canal. A canal can be constructed by dredging a channel in the bottom of an existing lake. When the channel is complete, the lake is drained and the channel becomes a new canal, serving both drainage of the surrounding polder and providing transport there. Examples include the Lage Vaart (nl). One can also build two parallel dikes in an existing lake, forming the new canal in between, and then drain the remaining parts of the lake. The eastern and central parts of the North Sea Canal were constructed in this way. In both cases pumping stations are required to keep the land surrounding the canal dry, either pumping water from the canal into surrounding waters, or pumping it from the land into the canal.

Canalization and navigations

A stream can be canalized to make its navigable path more predictable and easier to maneuver. Canalization modifies the stream to carry traffic more safely by controlling the flow of the stream by dredging, damming and modifying its path. This frequently includes the incorporation of locks and spillways, that make the river a navigation. Examples include the Lehigh Canal in Northeastern Pennsylvania's coal Region, Basse Saône, Canal de Mines de Fer de la Moselle, and Aisne River. Riparian zone restoration may be required.

Lateral canals

When a stream is too difficult to modify with canalization, a second stream can be created next to or at least near the existing stream. This is called a lateral canal, and may meander in a large horshoe bend or series of curves some distance from the source waters stream bed lengthening the effective length in order to lower the ratio of rise over run (slope or pitch). The existing stream usually acts as the water source and the landscape around its banks provide a path for the new body. Examples include the Chesapeake and Ohio Canal, Canal latéral à la Loire, Garonne Lateral Canal and Juliana Canal.

Smaller transportation canals can carry barges or narrowboats, while ship canals allow seagoing ships to travel to an inland port (e.g., Manchester Ship Canal), or from one sea or ocean to another (e.g., Caledonian Canal, Panama Canal).

Features

A canal boat traverses the longest and highest aqueduct in the UK, at Pontcysyllte in Denbighshire, Wales.

At their simplest, canals consist of a trench filled with water. Depending on the stratum the canal passes through, it may be necessary to linethe cut with some form of watertight material such as clay or concrete. When this is done with clay, it is known as puddling.

The Corinth Canal seen from the air.

Canals need to be level, and while small irregularities in the lie of the land can be dealt with through cuttings and embankments, for larger deviations other approaches have been adopted. The most common is the pound lock, which consists of a chamber within which the water level can be raised or lowered connecting either two pieces of canal at a different level or the canal with a river or the sea. When there is a hill to be climbed, flights of many locks in short succession may be used.

Prior to the development of the pound lock in 984 AD in China by Chhaio Wei-Yo and later in Europe in the 15th century, either flash locksconsisting of a single gate were used or ramps, sometimes equipped with rollers, were used to change the level. Flash locks were only practical where there was plenty of water available.

Locks use a lot of water, so builders have adopted other approaches for situations where little water is available. These include boat lifts, such as the Falkirk Wheel, which use a caisson of water in which boats float while being moved between two levels; and inclined planeswhere a caisson is hauled up a steep railway.

To cross a stream, road or valley (where the delay caused by a flight of locks at either side would be unacceptable) the valley can be spanned by a navigable aqueduct - a famous example in Wales is the Pontcysyllte Aqueduct (now a UNESCO World Heritage Site) across the valley of the River Dee.

Another option for dealing with hills is to tunnel through them. An example of this approach is the Harecastle Tunnel on the Trent and Mersey Canal. Tunnels are only practical for smaller canals.

Some canals attempted to keep changes in level down to a minimum. These canals known as contour canals would take longer, winding routes, along which the land was a uniform altitude. Other, generally later, canals took more direct routes requiring the use of various methods to deal with the change in level.

Canals have various features to tackle the problem of water supply. In cases, like the Suez Canal, the canal is simply open to the sea. Where the canal is not at sea level, a number of approaches have been adopted. Taking water from existing rivers or springs was an option in some cases, sometimes supplemented by other methods to deal with seasonal variations in flow. Where such sources were unavailable, reservoirs—either separate from the canal or built into its course—and back pumping were used to provide the required water. In other cases, water pumped from mines was used to feed the canal. In certain cases, extensive "feeder canals" were built to bring water from sources located far from the canal.

Where large amounts of goods are loaded or unloaded such as at the end of a canal, a canal basin may be built. This would normally be a section of water wider than the general canal. In some cases, the canal basins contain wharfs and cranes to assist with movement of goods.

When a section of the canal needs to be sealed off so it can be drained for maintenance stop planks are frequently used. These consist of planks of wood placed across the canal to form a dam. They are generally placed in pre-existing grooves in the canal bank. On more modern canals, "guard locks" or gates were sometimes placed to allow a section of the canal to be quickly closed off, either for maintenance, or to prevent a major loss of water due to a canal breach.

History

The transport capacity of pack animals and carts is limited. A mule can carry an eighth-ton [250 pounds (113 kg)] maximum load over a journey measured in days and weeks, though much more for shorter distances and periods with appropriate rest. Besides, carts need roads. Transport over water is much more efficient and cost-effective for large cargoes. It goes back to the earliest days of recorded history.

Ancient canals

The oldest known canals were irrigation canals, built in Mesopotamia circa 4000 BC, in what is now Iraq and Iran. The Indus Valley Civilization, Ancient India, (circa 2600 BC) had sophisticated irrigation and storage systems developed, including the reservoirs built at Girnar in 3000 BC. In Egypt, canals date back at least to the time of Pepi I Meryre (reigned 2332–2283 BC), who ordered a canal built to bypass the cataract on the Nile near Aswan.

In ancient China, large canals for river transport were established as far back as the Warring States period (481–221 BC), the longest one of that period being the Hong Gou (Canal of the Wild Geese), which according to the ancient historian Sima Qian connected the old states of Song, Zhang, Chen, Cai, Cao, and Wei. By far the longest canal was the Grand Canal of China, still the longest canal in the world today, and the oldest extant one. It is 1,794 kilometres (1,115 mi) long and was built to carry the Emperor Yang Guang between Zhuodu (Beijing) and Yuhang (Hangzhou). The project began in 605 and was completed in 609, although much of the work combined older canals, the oldest section of the canal existing since at least 486 BC. Even in its narrowest urban sections it is rarely less than 30 metres (98 ft) wide.

Greek engineers were the first to use canal locks, by which they regulated the water flow in the Ancient Suez Canal as early as the 3rd century BC.

"There was little experience moving bulk loads by carts, while a packhorse would [i.e. 'could'] carry only an eighth of a ton. On a soft road a horse might be able to draw 5/8ths of a ton. But if the load were carried by a barge on a waterway, then up to 30 tons could be drawn by the same horse

—technology historian Ronald W. Clark referring to transport realities before the industrial revolution and the Canal age.."

Middle Ages

In the Middle Ages, water transport was several times cheaper and faster than transport overland. Overland transport by animal drawnconveyances was used around settled areas, but unimproved roads required pack animal trains, usually of mules to carry any degree of mass, and while a mule could carry an eighth ton, it also needed teamsters to tend it and one man could only tend perhaps five mules,meaning overland bulk transport was also expensive, as men expect compensation in the form of wages, room and board. This was because long-haul roads were unpaved, more often than not too narrow for carts, much less wagons, and in poor condition, wending their way through forests, marshy or muddy quagmires as often as unimproved but dry footing. In that era, as today, greater cargoes, especially bulk goods and raw materials, could be transported by ship far more economically than by land; in the pre-railroad days of the industrial revolution, water transport was the gold standard of fast transportation. The first artificial canal in Western Europe was the Fossa Carolinabuilt at the end of the 8th century under personal supervision of Charlemagne.

In Britain, the Glastonbury Canal is believed to be the first post-Roman canal and was built in the middle of the 10th century to link the River Brue at Northover with Glastonbury Abbey, a distance of about 1.75 kilometres (1,900 yd). Its initial purpose is believed to be the transport of building stone for the abbey, but later it was used for delivering produce, including grain, wine and fish, from the abbey's outlying properties. It remained in use until at least the 14th century, but possibly as late as the mid-16th century.

More lasting and of more economic impact were canals like the Naviglio Grande built between 1127 and 1257 to connect Milan with the Ticino River. The Naviglio Grande is the most important of the lombard "navigli" and the oldest functioning canal in Europe.

Later, canals were built in the Netherlands and Flanders to drain the polders and assist transportation of goods and people.

Canal building was revived in this age because of commercial expansion from the 12th century. River navigations were improved progressively by the use of single, or flash locks. Taking boats through these used large amounts of water leading to conflicts with watermill owners and to correct this, the pound or chamber lock first appeared, in the 10th century in China and in Europe in 1373 in Vreeswijk, Netherlands. Another important development was the mitre gate, which was, it is presumed, introduced in Italy by Bertola da Novate in the

16th century. This allowed wider gates and also removed the height restriction of guillotine locks.

To break out of the limitations caused by river valleys, the first summit level canals were developed with the Grand Canal of China in 581–617 AD whilst in Europe the first, also using single locks, was the Stecknitz Canal in Germany in 1398.

Early modern period

c. 1500–1800 The first canal to use pound locks was the Briare Canal connecting the Loire and Seine (1642), followed by the more ambitious Canal du Midi (1683) connecting the Atlantic to the Mediterranean. This included a staircase of 8 locks at Béziers, a 157 metres (515 ft) tunnel and three major aqueducts.

Canal building progressed steadily in Germany in the 17th and 18th centuries with three great rivers, the Elbe, Oder and Weser being linked by canals. In post-Roman Britain, the first early modern period canal built appears to have been the Exeter Canal, which was surveyed in 1563, and open in 1566.

The oldest canal, technically a mill race built for industrial purposes in North America is Mother Brook, also known as mill brook, in between the two Boston, Massachusetts neighbourhoods of Dedham and Hyde Park connecting the higher waters of the Charles River and the mouth of the Neponset River and the sea. It was constructed in 1639 to provide water power for mills.

In Russia, the Volga–Baltic Waterway, a nationwide canal system connecting the Baltic Sea and Caspian Sea via the Neva and Volgarivers, was opened in 1718.

INDUSTRIAL REVOLUTION

The modern canal system was mainly a product of the 18th century and early 19th century. It came into being because the Industrial Revolution (which began in Britain during the mid-18th century) demanded an economic and reliable way to transport goods and commodities in large quantities.

By the early 18th century, river navigations such as the Aire and Calder Navigation were becoming quite sophisticated, with pound locksand longer and longer "cuts" (some with intermediate locks) to avoid circuitous or difficult stretches of river. Eventually, the experience of building long multi-level cuts with their own locks gave rise to the idea of building a "pure" canal, a waterway designed on the basis of where goods needed to go, not where a river happened to be.

The claim for the first pure canal in Great Britain is debated between "Sankey" and "Bridgewater" supporters. The first true canal in what is now the United Kingdom was the Newry Canal in Northern Ireland constructed by Thomas Steers in 1741.

The Sankey Brook Navigation, which connected St Helens with the River Mersey, is often claimed as the first modern "purely artificial" canal because although originally a scheme to make the Sankey Brook navigable, it included an entirely new artificial channel that was effectively a canal along the Sankey Brook valley. However, "Bridgewater" supporters point out that the last quarter-mile of the navigation is indeed a canalized stretch of the Brook, and that it was the Bridgewater Canal (less obviously associated with an existing river) that captured the popular imagination and inspired further canals.

In the mid-eighteenth century the 3rd Duke of Bridgewater, who owned a number of coal mines in northern England, wanted a reliable way to transport his coal to the rapidly industrializing city of Manchester. He commissioned the engineer James Brindley to build a canal for that purpose. Brindley's design included an aqueduct carrying the canal over the River Irwell. This was an engineering wonder which immediately attracted tourists. The construction of this canal was funded entirely by the Duke and was called the Bridgewater Canal. It opened in 1761 and was the first major British canal.

The new canals proved highly successful. The boats on the canal were horse-drawn with a towpath alongside the canal for the horse to walk along. This horse-drawn system proved to be highly economical and became standard across the British canal network. Commercial horse-drawn canal boats could be seen on the UK's canals until as late as the 1950s, although by then diesel-powered boats, often towing a second unpowered boat, had become standard.

The canal boats could carry thirty tons at a time with only one horse pulling - more than ten times the amount of cargo per horse that was possible with a cart. Because of this huge increase in supply, the Bridgewater canal reduced the price of coal in Manchester by nearly two-thirds within just a year of its opening. The Bridgewater was also a huge financial success, with it earning what had been spent on its construction within just a few years.

This success proved the viability of canal transport, and soon industrialists in many other parts of the country wanted canals. After the Bridgewater canal, early canals were built by groups of private individuals with an interest in improving communications. In Staffordshire the famous potter Josiah Wedgwood saw an opportunity to bring bulky cargoes of clay to his factory doors and to transport his fragile finished goods to market in Manchester, Birmingham or further away, by water, minimizing breakages. Within just a few years of the Bridgewater's opening, an embryonic national canal network came into being, with the construction of canals such as the Oxford Canal and the Trent & Mersey Canal.

The new canal system was both cause and effect of the rapid industrialization of The Midlands and the north. The period between the 1770s and the 1830s is often referred to as the "Golden Age" of British canals.

For each canal, an Act of Parliament was necessary to authorize construction, and as people saw the high incomes achieved from canal tolls, canal proposals came to be put forward by investors interested in profiting from dividends, at least as much as by people whose businesses would profit from cheaper transport of raw materials and finished goods.

In a further development, there was often out-and-out speculation, where people would try to buy shares in a newly floated company simply to sell them on for an immediate profit, regardless of whether the canal was ever profitable, or even built. During this period of "canal mania", huge sums were invested in canal building, and although many schemes came to nothing, the canal system rapidly expanded to nearly 4,000 miles (over 6,400 kilometres) in length.

Many rival canal companies were formed and competition was rampant. Perhaps the best example was Worcester Bar in Birmingham, a point where the Worcester and Birmingham Canal and the Birmingham Canal Navigations Main Line were only seven feet apart. For many years, a dispute about tolls meant that goods travelling through Birmingham had to be portaged from boats in one canal to boats in the other.

Canal companies were initially chartered by individual states in the United States. These early canals were constructed, owned, and operated by private joint-stock companies. Three were completed when the War of 1812 broke out; these were the Santee Canal (opened 1800) in South Carolina, the Middlesex Canal (opened 1802) in Massachusetts and the Dismal Swamp Canal (opened 1805) in Virginia. The Erie Canal (opened 1825) was chartered and owned by the state of New York and financed by bonds bought by private investors. The Erie canal runs about 363 miles (584 km) from Albany, New York, on the Hudson River to Buffalo, New York, at Lake Erie. The Hudson River connects Albany to the Atlantic port of New York City and the Erie Canal completed a navigable water route from the Atlantic Ocean to the Great Lakes. The canal contains 36 locks and encompasses a total elevation differential of around 565 ft. (169 m). The Erie Canal with its easy connections to most of the U.S. mid-west and New York City soon quickly paid back all its invested capital (US$7 million) and started turning a profit. By cutting transportation costs in half or more it became a large profit centre for Albany and New York City as it allowed the cheap transportation of many of the agricultural products grown in the mid west of the United States to the rest of the world. From New York City these agricultural products could easily be shipped to other U.S. states or overseas. Assured of a market for their farm products the settlement of the U.S. mid-west was greatly accelerated by the Erie Canal. The profits generated by the Erie Canal project started a canal building boom in the United States that lasted until about 1850 when railroads started becoming seriously competitive in price and convenience. The Blackstone Canal (finished in 1828) in Massachusetts and Rhode Island fulfilled a similar role in the early industrial revolution between 1828 and 1848. The Blackstone Valley was a major contributor of the American Industrial Revolution where Samuel Slater built his first textile mill.

Power canals

A power canal refers to a canal used for hydraulic power generation, rather than for transport. Nowadays power canals are built almost exclusively as parts of hydroelectric power stations. Parts of the United States, particularly in the Northeast, had enough fast-flowing rivers that water power was the primary means of powering factories (usually textile mills) until after the American Civil War. For example, Lowell, Massachusetts, considered to be "The Cradle of the American Industrial Revolution," has 6 miles (9.7 km) of canals, built from around 1790 to 1850, that provided water power and a means of transportation for the city. The output of the system is estimated at 10,000 horsepower. Other cities with extensive power canal systems include Lawrence, Massachusetts, Holyoke, Massachusetts, Manchester, New Hampshire, and Augusta, Georgia. The most notable power canal was built in 1862 for the Niagara Falls Hydraulic Power and Manufacturing Company.

19th century

Competition, from railways from the 1830s and roads in the 20th century, made the smaller canals obsolete for most commercial transport, and many of the British canals fell into decay. Only the Manchester Ship Canal and the Aire and Calder Canal bucked this trend. Yet in other countries canals grew in size as construction techniques improved. During the 19th century in the US, the length of canals grew from 100 miles (161 km) to over 4,000, with a complex network making the Great Lakes navigable, in conjunction with Canada, although some canals were later drained and used as railroad rights-of-way.

In the United States, navigable canals reached into isolated areas and brought them in touch with the world beyond. By 1825 the Erie Canal, 363 miles (584 km) long with 36 locks, opened up a connection from the populated Northeast to the Great Lakes. Settlers flooded into regions serviced by such canals, since access to markets was available. The Erie Canal (as well as other canals) was instrumental in lowering the differences in commodity prices between these various markets across America. The canals caused price convergence between different regions because of their reduction in transportation costs, which allowed Americans to ship and buy goods from farther distances much cheaper. Ohio built many miles of canal, Indiana had working canals for a few decades, and the Illinois and Michigan Canalconnected the Great Lakes to the Mississippi River system until replaced by a channelized river waterway.

Three major canals with very different purposes were built in what is now Canada. The first Welland Canal, which opened in 1829 between Lake Ontario and Lake Erie, bypassing Niagara Falls and the Lachine Canal (1825), which allowed ships to skirt the nearly impassable rapids on the St. Lawrence River at Montreal, were built for commerce. The Rideau Canal, completed in 1832, connects Ottawa on the Ottawa River to Kingston, Ontario on Lake Ontario. The Rideau Canal was built as a result of the War of 1812 to provide military transportation between the British colonies of Upper Canada and Lower Canada as an alternative to part of the St. Lawrence River, which was susceptible to blockade by the United States.

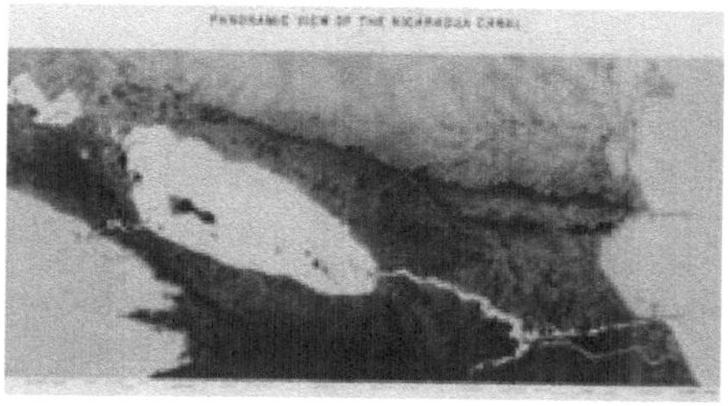

In France, a steady linking of all the river systems — Rhine, Rhône, Saône and Seine — and the North Sea was boosted in 1879 by the establishment of the Freycinet gauge, which specified the minimum size of locks. Canal traffic doubled in the first decades of the 20th century.

Many notable sea canals were completed in this period, starting with the Suez Canal (1869) - which carries tonnage many times that of most other canals - and the Kiel Canal (1897), though the Panama Canal was not opened until 1914.

In the 19th century, a number of canals were built in Japan including the Biwako canal and the Tone canal. These canals were partially built with the help of engineers from the Netherlands and other countries.

Modern uses

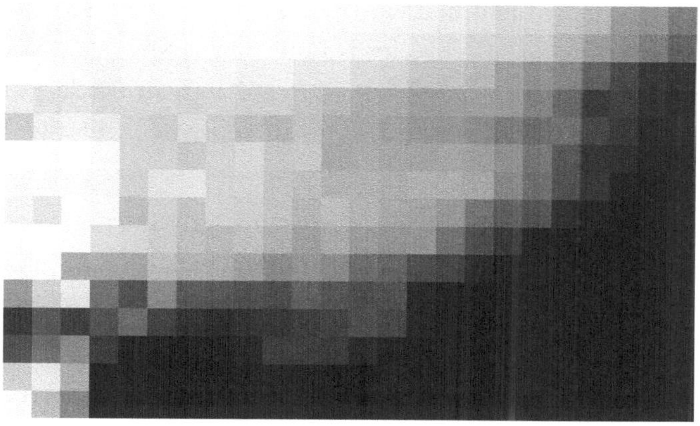

Large-scale ship canals such as the Panama Canal and Suez Canal continue to operate for cargo transportation, as do European barge canals. Due to globalization, they are becoming increasingly important, resulting in expansion projects such as the Panama Canal expansion project. The expanded canal began commercial operation on 26 June 2016. The new set of locks allow transit of larger, Post-Panamax and New Panamax ships.

The narrow early industrial canals, however, have ceased to carry significant amounts of trade and many have been abandoned to navigation, but may still be used as a system for transportation of untreated water. In some cases railways have been built along the canal route, an example being the Croydon Canal.

A movement that began in Britain and France to use the early industrial canals for pleasure boats, such as hotel barges, has spurred rehabilitation of stretches of historic canals. In some cases, abandoned canals such as the Kennet and Avon Canal have been restored and are now used by pleasure boaters. In Britain, canalside housing has also proven popular in recent years.

The Seine–Nord Europe Canal is being developed into a major transportation waterway, linking France with Belgium, Germany, and the Netherlands.

Canals have found another use in the 21st century, as easements for the installation of fibre optic telecommunications network cabling, avoiding having them buried in roadways while facilitating access and reducing the hazard of being damaged from digging equipment.

Canals are still used to provide water for agriculture. An extensive canal system exists within the Imperial Valley in the Southern California desert to provide irrigation to agriculture within the area.

Canals are so deeply identified with Venice that many canal cities have been nicknamed "the Venice of...". The city is built on marshy islands, with wooden piles supporting the buildings, so that the land is man-made rather than the waterways. The islands have a long history of settlement; by the 12th century, Venice was a powerful city state.

Amsterdam was built in a similar way, with buildings on wooden piles. It became a city around 1300. Many Amsterdam canals were built as part of fortifications. They became grachten when the city was enlarged and houses were built alongside the water.

Other cities with extensive canal networks include: Alkmaar, Amersfoort, Bolsward, Brielle, Delft, Den Bosch, Dokkum, Dordrecht, Enkhuizen, Franeker, Gouda, Haarlem, Harlingen, Leeuwarden, Leiden, Sneek and Utrecht in the Netherlands; Brugge and Gent in Flanders, Belgium; Birminghamin England; Saint Petersburg in Russia; Aveiro in Portugal; Hamburg and Berlin in Germany; Fort Lauderdale and Cape Coral in Florida, United States and Lahore in Pakistan.

Liverpool Maritime Mercantile City is a UNESCO World Heritage Site near the centre of Liverpool, England, where a system of intertwining waterways and docks is now being developed for mainly residential and leisure use.

Canal Estates (commonly known as bayous) are a form of subdivision popular in cities like Miami, Florida, Texas City, Texas and the Gold Coast, Queensland; the Gold Coast has over 700 km of residential canals. Wetlands are difficult areas upon which to build housing estates, so dredging part of the wetland down to a navigable channel provides fill to build up another part of the wetland above the flood level for houses. Land is built up in a finger pattern that provides a suburban street layout of waterfront housing blocks.

Boats

Inland canals have often had boats specifically built for them. An example of this is the British narrowboat, which is up to 72 feet (21.95 m) long and 7 feet (2.13 m) wide and was primarily built for British Midland canals. In this case the limiting factor was the size of the locks. This is also the limiting factor on the Panama canal where Panamax ships were limited to a length of 289.56 m (950 ft) and a beam of 32.31 m (106 ft) until 26 June 2016 when the opening of larger locks allowed for the passage of larger New Panamax ships. For the lockless Suez Canal the limiting factor for Suezmax ships is generally draft, which is limited to 16 m (52.5 ft). At the other end of the scale, tub-boat canals such as the Bude Canal were limited to boats of under 10 tons for much of their length due to the capacity of their inclined planes or boat lifts. Most canals have a limit on height imposed either by bridges or by tunnels.

Canal

Canals are waterway channels, or artificial waterways, for water conveyance, or for servicing water transport vehicles. They carry free surface flow under atmospheric pressure and can be thought of as artificial rivers.

Canals are a human-made waterway that allows boats and ships to pass from one body of water to another. Canals are also used to transport water for irrigation and other human uses.

There are two types of canals: waterways and aqueducts. Waterways are the navigable parts of a body of water and can be located within a bay or open sea, can connect two or more water bodies, or may even form networks within a city.

Aqueducts are used exclusively to transport water for drinking, agriculture, and hydroelectric power.

Canals, channels, and reservoirs constructed, maintained, or controlled by the State Government for the supply or storage of water.

List of Important Canals in India

Canal Name	Area Served
Sethusamudram Shipping Canal Project	Tamilnadu, Kerala
Indira Gandhi Canal	Sultanpur, Punjab, Rajasthan
Buckingham Canal	Andhra Pradesh, Tamil Nadu
National Waterway 3 (India) (West Coast Canal)	Kerala
Upper Ganges Canal	Uttrakhand, Uttar Pradesh
Lower Ganges Canal	Uttrakhand, Uttar Pradesh
Agra Canal	Uttar Pradesh, Haryana, Rajasthan
Telugu Ganga project	Andhra Pradesh
Handri-Neeva	Andhra Pradesh
K. C. Canal	Andhra Pradesh
Tajewala Canal	Haryana
Munak canal (part of Western Yamuna Canal)	Haryana & Delhi
Western Yamuna Canal	Haryana & Delhi
Kalingarayan Canal	Tamilnadu
Buddha Nullah Canal	Punjab
Sutlej Yamuna link Canal	Punjab
Sirhind Canal	Punjab
Anupgarh canal	Rajasthan (western India)
Narmada Canal	Gujarat, Rajasthan
Conolly Canal	Kerala
Ainsley canal	Tamilnadu
Kalingarayan Canal	Tamilnadu
Basavarajeshwari Camp	Karnataka
Lower Bhavani Project Canal	Tamilnadu
Kaveri-Vaigai link canal	Kerala, Karnataka, Tamilnadu
Trivandrum-Shoranur canal	Kerala
Tamirabarani-Nambiar link canal	Kerala, Karnataka
Kakatiya Canal	Telangana
Sharda Canal	Uttar Pradesh
Zamania Canal	Uttar Pradesh
Deokali Canal	Uttar Pradesh

Karmanasa Canal	Bihar
Durgavati Canal	Bihar
Kutte Kol	Jammu and Kashmir
Nallah Mar or Mar Canal	Jammu and Kashmir
Tsoont Kol	Jammu and Kashmir
Cumbarjua Canal	Goa
Soundane Cut	Maharashtra

8

CANAL BUILDING TECHNIQUE

Canals have to be designed and built to withstand enormous water pressure. In addition, they have to be watertight so that they will not leak and they have to be able offer protection from the risk of erosion caused by the flowing water.

The bottom of the canal is levelled off and compacted. A filter layer consisting of sand and gravel is then put in place and, on top of that, a sealing layer of fine soil, cement or asphalt. There is always a certain amount of leakage through the sealant and the filter ensures that the sealant does not get washed away. If fine soil is used as the sealing layer, it must be covered by an erosion resistant layer. It is highly important that the various layers are correctly compacted to avoid cracks in the sealing layer. Even before the Revolutionary War gave New impetus to American expansionism, the colonial political and economic elites were deeply interested in the improvement of inland transportation. Vessels that plied offshore waters, small boats and rafts on the streams down to tidewater, and local roads and turnpikes served the immediate commercial needs of farmers and townspeople in the Atlantic coastal area. But the loftier dreams of planters, merchants, and political leaders—as well as of the common farmers who constituted by far most of the free population in British America—looked beyond the "fall line" that separated the rivers flowing to the coast from those that ran to the Ohio-Mississippi basin. A vast area for settlement and productivity—and riches—lay in the interior, and by the early 1790s demands for diffusion of New transport technologies and for investment in internal improvements were voiced frequently in both state and national political forums.

It was widely recognized that unless bulk agricultural commodities, which were the staples of a commercialized and expanding farm economy, could be carried cheaply. And over long distances, settlement and economic growth would be badly hampered in the region beyond the Appalachian Mountains. Then, too, there were opportunities for construction of short lines on the Atlantic seaboard to link already developed areas (coal mines, farming and lumber regions, and rising industrial sites), with the promise of immediate traffic and revenues. The latter were "exploitative" projects, tapping existing trade routes and resources; but the major east-west projects were "developmental," promoted with the goal of opening Newly or sparsely settled areas to economic opportunities. There was also a nationalistic or patriotic goal of canal promotion: to bind together far-flung sections of the young nation and to prove the efficacy of republican government.

And yet total canal construction in the United States up to 1816 totaled only 100 miles—the longest canal project being the Middlesex, which linked Boston's harbor with the farm region to the north. Other lines of some importance linked Norfolk, Virginia, to Albemarle Sound and connected the Santee River area to Charleston, South Carolina. Although many other canal projects were proposed up and down the Atlantic coast, progress was difficult because of shortages of capital and skepticism with regard to engineering feasibility projects. Moreover, regional or local jealousies notoriously worked against successful mobilization of governmental support in both the U.S. Congress and the state legislatures.

In the period from the mid-1820s to the Civil War, however, the United States underwent a vast expansion of canal construction, becoming the world's leading nation in both mileage of canals and the volume of tonnage carried on them. The canal lines were of crucial importance in the integration of a national economy, and they played a key role in the so-called "Transportation Revolution" that expedited both westward expansion and a robust industrialization process in the North and West.

Advantages, Disadvantages, and Construction Challenges

Canal technology proved especially attractive for several reasons. Since the 1760s, successful large-scale canal projects had been built in both Great Britain and France, and these canals had brought enormous economic advantages to the regions they served. The engineering advances pioneered in Europe gave American promoters confidence that they could build canals with equal success.

There was a downside to canal technology, too, though it was not always fully recognized in America. Difficult topography or uncertain water supply meant complex and highly expensive construction design. Canal building before the

1850s was mainly done with hand tools, augmented only by some primitive animal-powered machinery. A canal line had to be furnished with locks, permitting boats to pass through from one water level to another. The segments of line between the locks were of very gradual grade to permit controlled flow. At each lock, a gate at its higher level would be opened while the gate on the lower end was kept closed; once a boat entered the lock's chamber from the higher level, the upper gate was closed (holding the water flow back) while the lower gate was opened. As water ran out, the boat was carried down to the lower level, then passed through the open gate. For "upstream" movement, from the lower level to the higher, the process was reversed. The lock would be drained to the lower level, the boat would enter through the bottom gate, which was then closed, and water would then be admitted from the upper gate, lifting the boat up to the higher level. In steep areas, "flights" of locks, closely spaced, were necessary and often involved complex engineering; for transit of the boats, these series of locks meant a slow stretch and usually long waiting periods.

Locks varied in size. Lifts ranged from two to thirty feet, and there were great differences in the distances between gates as well as in the construction materials used. Masonry locks and metal or metal-trimmed gates were far more expensive—and more durable—than wooden gates and timber-supported rubble for the walls as found on some of the lines. The total rise and fall over an entire line was measured as "lockage," and served as an index of the difficulty of construction. For example, New York's Erie Canal route measured lockage of 655 feet, by contrast with Pennsylvania's lockage of 3,358 feet between Philadelphia and the Ohio River.

The size of the boats that could be accommodated, as well as the volume of water needed, were functions of the dimensions of the canal bed, or its "prism," as well as of the size of lock chambers. Prisms on American canals varied greatly, most of them ranging from forty to sixty feet in width at the top, with sloping sidewalls leading to a bottom of twenty-five to forty-five feet across. The Pennsylvania system was the most complex in engineering using inclined planes and steam-powered winches to drag boats out of the water and over some of the steepest hills.

To supply the line with flowing water, engineering plans had to include river connections, dams and reservoirs with feeder lines to the canals, and often massive culverts and aqueducts. Building the sidewalls to minimize loss of water through seepage was another challenging and expensive aspect of design. On many of the larger canals, such as the Ohio lines, engineers took advantage of fast-flowing feeder streams to design water-mill sites into the line. Once a canal was in operation, moreover, maintaining navigation was a continuous challenge. Winter ice, droughts, floods, and breaches in the water-supply system would frequently cause navigational closings. Even in the best of circumstances, it was difficult to maintain regular schedules on the lines because of traffic bottlenecks at the locks and the continuous maintenance needed to keep water flowing.

Although steam-powered propeller craft were used on a few canal lines, this form of transport placed dangerous pressure on the canal walls. Hence, the use of horses or mules to haul canal boats was nearly universal, with the animals walking along the "towpath" alongside the line. Freight boats typically of 50-to 125-ton capacity operated at speeds of one to three miles per hour. On most lines they were owned by individuals or private companies, the line being a common carrier under the law.

In the short run, all the disadvantages of canal technology were more than offset by the cost savings for long-distance hauling, especially of bulk goods and produce. In the long run, however, innovations in steam technology and railroad engineering were destined to render many canals the losers in a New competitive age in transport that took shape in the late 1840s and the 1850s.

The Erie Canal

The great breakthrough came in 1817 with New York State's commitment to building the Erie Canal to connect the Hudson River at Albany with Buffalo on Lake Erie, a Project far greater than any previously attempted in America. The Erie was important to subsequent canal development in several ways, most notably because it provided a model of public enterprise through its financing, administration, and implementation. The state raised capital through bond issues both in New York and in Europe, and supplemented these funds with tax revenues. Actual construction was overseen by a board of commissioners, some of whom were personally involved in the fieldwork, but the project became a celebrated "school for engineers," with most of the junior personnel learning their skills on the job under the tutelage of Benjamin Wright and James Geddes, two of less than twenty men who then constituted the profession in America. Many of the Erie engineers went on to direct canal surveys in other states.

The canal was divided into sections for purpose of construction, with private contractors taking on the work under the state engineers' supervision—a scheme that was emulated by nearly all the major canals subsequently built. It was an immediate commercial success once opened to its full length in 1825, leading the New York legislature to authorize a series of additional canals as well as the improvement and enlargement of the original line.

No nonmilitary enterprise in the United States had ever involved such expenditures as did the Erie, whose initial construction cost $6 million. The number of laborers employed was also unprecedented in any economic enterprise of the day. The state's construction expenditures energized local economies, giving part-time employment to farmers and creating sudden demand for stone, timber, mules, and oxen, and provisions for workers. Like canals and other public works throughout the country, moreover, the Erie attracted immigrant workers (mainly Irish and German) who were employed to do much of the most dangerous work.

The Erie's commercial impact on the rural countryside and on New York City's role as a center for trade with the interior and for exports to Europe—together with the rich stream of revenues from the tolls—heightened expectations everywhere in the country that other canals could produce equally spectacular fiscal and developmental results.

The Post-1825 Boom in Canal Building

Emulation of New York followed quickly. In 1825 Pennsylvania authorized a $10 million project, combining canal technology with the use of inclined planes. It was completed in 1834, tapping the Ohio Valley's farm country at Pittsburgh and giving Philadelphia trade advantages similar to those that its rival New York City had obtained from the Erie. The first of the western states to build a major line was Ohio, which authorized construction in 1825. Although still small in population and financial resources, Ohio, too, resorted to creation of a state enterprise and borrowed heavily both in the East and in Europe. Erie Canal engineers were brought in at first, but Alfred Kelley of Cleveland and Micajah Williams of Cincinnati,

Local entrepreneurs with no prior engineering experience, took principal charge of overseeing construction once the technical plans were adopted. Although administrative incompetence and corruption plagued the Pennsylvania project, Ohio's record was widely admired for its efficiency and strength of design. One line, the Ohio Canal, was completed in 1834 and extended from Cleveland on Lake Erie to Portsmouth on the Ohio River—the first water link between the Great Lakes and the great Mississippi-Ohio basin. A second line, completed in the mid-1840s, linked Cincinnati with Toledo to the north.

Other important lines begun or fully built prior to 1840 included the Delaware and Hudson Canal, a successful private line in the Pennsylvania coal country; the Delaware and Raritan Canal, also private, linking Philadelphia and

New York; and the Chesapeake and Delaware Ohio Canal, which with substantial state support built a line, surveyed by the engineer William Strickland, through Maryland to link Baltimore with the Philadelphia port.

In the period 1815–1834, $60 million was invested in 2,088 miles of canals, with 70 percent of the funds coming from governmental sources, mainly the states. Most of the funds were borrowed at home and abroad. Also, Congress authorized the Army Engineers to conduct surveys for the states and federal companies; made some direct federal investments; and gave several million acres of public lands to Ohio, Indiana, and Illinois to subsidize their canal projects during this period.

In the decade following, 1834–1844, the "canal enthusiasm" continued to animate state governments and private promoters. Rivalries among states and competition among cities were intense, feeding the spirit of optimism. A new wave of canal construction followed, with the projects again heavily financed by loans from Europe and the eastern cities. Almost 1,300 miles of canal were built during this ten-year period. They cost $72 million, of which 79 percent represented public funds. In addition to major New state canal systems begun in Illinois and in Indiana (where the Wabash and Erie line would open another link for direct trade between the Ohio River and Lake Erie), and in Illinois, three of the pioneering state projects—the Erie, Ohio's two main canals, and the Pennsylvania system—were further expanded to satisfy sections of their states that had been left out of the original system designs. As the New canals were generally of larger dimensions than the first ones to be built, the carrying capacity for canal traffic doubled between 1834 and 1844. Until 1839, conditions of prosperity and expansion sustained the canal-building movement, and expenditures for the New canals stimulated overall economic growth.

Financial Problems and Railroad Competition

The 1837 financial panic and the 1839–1843 depression created enormous fiscal problems for many canal states, leading to defaults on state debts in Pennsylvania and Indiana. Because many of the expansion projects and new lines produced toll revenues far below expectations, moreover, there was widespread disillusionment with state enterprise; and this became a factor favoring railroads as an alternative to canals, especially given the much greater reliability of rail transport. In the Ohio-Indiana-Illinois area, by 1848 the proliferation of canal lines also produced intensified competition between the various Great Lakes and Mississippi River routes, now also served by steamboat lines on these connecting waters. The east-west and local railroads of the 1850s made matters worse.

The result was heavy downward pressure on canal rates, consequently reduced revenues, and, soon, a scenario of operating deficits that placed an unwelcome burden on taxpayers.

Transport competition drove down rates so much that the period from the mid-1840s to the Civil War formed a distinctive "second phase" of the Transportation Revolution. By 1850–1852, for example, western canal tolls were less than a third the level of the 1830s, creating still further fiscal problems for the canal states and companies. Where private investment had been invited on a matching basis for "mixed" canal enterprises, the costs fell hard on the capitalists as well. But while revenues fell, ton-miles of canal transportation continued to expand on all the major lines throughout the 1850s.

During the period 1844–1860, a last major cycle of canal construction produced 894 miles of line at a cost of $57 million. Here again, governmental activism was crucial, with public funds accounting for two-thirds of the total expended. Much of this increase constituted the completion or improvement of lines built earlier; in addition, the still-successful Erie system in New York was further enlarged and upgraded. A large expenditure was made, too, on the Sault Ste. Marie Ship Canal, a short but massive deepwater project that connected Lake Huron with Lake Superior.

Although much of the canal system experienced operating deficits in the 1850s, the impetus these New facilities had given the economy had clearly warranted most of the capital invested. Commercialization of agriculture in the western states and other interior had been made possible, while eastern manufacturers and importers were afforded economical access to interior markets. Coal-mining and iron centers were linked, and consumer prices fell where the transport facilities had proliferated. In sum, the areas served by canals were enabled to build on comparative economic advantage; and, at least in the northern states, processing of primary products carried by the canals served as the origin of manufacturing growth that augmented urban commercial activity.

Railroad competition led to many closings of once-important canals; indeed, more than 300 miles of line were abandoned by 1860. A few of the canals did continue to carry heavy traffic after the Civil War. The most important to commerce in the twentieth century was the Atlantic intra-coastal waterway, which permitted vessels to transit offshore waters safely from New England to Florida. The Erie retained importance as a barge canal, as did some of the shorter coal-carrying lines. Some of the old canal lines became rights-of-way for railroads or modern roads; others were absorbed into the changing landscape as development went forward. In scattered locations, a few segments of the great canal lines are today preserved or restored for enjoyment of citizens seeking a glimpse of the once-glorious era of canal transport in America.

Prehistoric Engineering

Nobody in the New World moved water more efficiently than the Hohokam Indians. The canals they built to sustain their large and complex agricultural society are still considered an engineering marvel.

The Hohokam people lived around the Mesa area of Central Arizona for almost 1,500 years. They appeared around 1 CE and disappeared about 1450. Their agricultural society flourished around the middle Gila River and Salt River. These sources allowed them to farm beans, squash, corn, and cotton. The Hohokam were skilled agrarians, managing the soil to replace lost nutrients. Additional food came from extensive gathering of wild plants, dry farming agave, and hunting deer and small animals.

Around 600 CE the Hohokam people began to construct irrigation canals. They excavated trenches up to 12 feet deep by hand, using a digging stick. The trenches drew off Salt River water and fanned into a network of smaller canals that brought a steady supply of water to fields. Between 1100 CE and 1450 CE, 500 miles of canals irrigated 110,000 acres. The food produced by this advanced irrigation system is believed to have supported up to 80,000 people—the highest population density in the prehistoric Southwest.

Hohokam canals possessed remarkable features that still stand as examples of engineering genius. First, they were wide at the mouth and carefully tapered, getting smaller as secondary branches drew water from the main channel. By shrinking the channel size as the flow decreases, the Hohokam were able to stabilize flow rate. A steady flow rate is key to creating a functional irrigation system. Water that moves too fast carries sand and silt that will eventually block the canal. Water that moves too slowly will not reach its destination. Second, canals covered miles of territory on a relatively steady grade, avoiding hills and valleys that could affect flow rate.

Though the Hohokam people are most celebrated for their canals, they made fire-hardened pottery and figures and jewelry from shell and turquoise. They built increasingly elaborate structures, such as the Big House in Casa Grande Ruins National Monument. The society also constructed large earthen platform mounds of unconfirmed purpose. Located in the center of villages, the mounds may have been used for housing elite members of the society or for ceremonial activities. Hohokam rock art, in the form of petroglyphs and pictographs, is found in mountain ranges and rocky areas throughout the region.

MYSTERIOUS MIGRATION

The Hohokam disappeared from their villages near the Salt and Gila Rivers in 1450 CE. Their descendants survive as members of the Tohono O'odham Nation.

The three centuries leading up to the Hohokam's disappearance around 1450 CE had been a period of migration and social change, with old settlements and prestigious structures constructed in the Salt and Gila Basins being abandoned. The reasons for this mass abandonment remain unknown to this day. Perhaps disease or war threatened the population. Perhaps flood damaged the canal infrastructure, or drought brought hardship. There is also speculation that a great leader died, causing the people to drift away and migrate. We may never discover why the Hohokam decided to leave the area they had lived in for almost 1,500 years. But their descendants live on as members of the Tohono O'odham Nation.

Reportedly, canals were built in Egypt as much as 6,000 years ago. The first canal designed to accommodate seaworthy ships was built by the Pharaoh Sesosteris I4,000 years ago, linking the Red Sea to the Nile River and thus to the Mediterranean Sea. Remnants of this canal, which was alternately neglected and rebuilt half a dozen times, suggest that it was about 150 ft (46 m) wide, 16 ft (5 m) deep, and 60 mi (97 km) long.

The oldest canal still in use was begun in China in the fourth century B.C. for the purposes of collecting grain taxes and transporting troops. Rebuilt and enlarged about A.D. 600, the channel became known as the Grand Canal; its latest extension, completed about A.D. 1280, brought its length to 1,114 mi (1,795 km). It remains the longest navigable canal in the world.

For centuries, canals were dug using only hand tools. While building the Canal du Midi between 1665 and 1681 to link the Atlantic

An aerial view of a canal and lock system.

Ocean and the Mediterranean Sea across southern France, gunpowder was first used to blast a tunnel. The invention of steam-powered machinery at the end of the seventeenth century helped fuel the "canal eras" in Europe and North America in the eighteenth and nineteenth centuries during which thousands of miles of canals were built. Horse-drawn plows continued to be used especially for final trimming of the channel to desired slopes and depths. Full mechanization of canal construction was achieved in 1946, when an American company built the first canal trimming and lining machines.

Locks

Flash locks, the first attempts to carry boats over difficult elevation changes on rivers or canals, date from the third century B.C. Greek historian Diodorus Siculus, writing 200 years later, described how Ptolemy II had improved the Nile-to-Red Sea canal by building a type of lock. Consisting of a single gate, flash locks carried boats downstream on a rush of water; boats headed upstream could be pulled forward while floating on the torrent released by opening the gate.

Flash locks were hazardous, and they used large amounts of water. A significant improvement came in A.D. 984, when the first double-gate lock was built on China's Grand Canal. Also called a pound lock (because it impounds water) this was the predecessor of modern conventional locks. Its gates were panels that lifted vertically. A similar lock with vertical-lift gates built in the Netherlands in 1373 also controlled its water level by partially opening either the upstream or the downstream gate. Pound lock operation was greatly improved in 1485 when an Italian lock was built with smaller, valve-controlled openings in the gates.

Leonardo da Vinci invented the miter gate in 1480. Two gates, each more than half as wide as the lock, swing on vertical hinges. In the open position, they are flush with the lock's walls. In the closed position, they meet in a V pointed upstream so that the higher water level presses against them to promote a tight seal.

The base of the upstream gate in a lock is higher than the base of the downstream gate. If the lock was very deep, water filling the chamber through an opening in the upper gate would create turbulence or even swamp a boat in the chamber. This problem was solved in France during the seventeenth century, when valve-controlled water channels for filling and emptying a lock were built

A lock is a rectangular chamber with gates at both ends; with both gates closed, the water level within the lock can be adjusted to match the canal water level on either side. Thus, a vessel entering the lock can be raised or lowered in order to enter the next level canal section.

Until the early nineteenth century, lock gates were made of wood, and lock chambers were lined with wood, stone, bricks, or turf. In 1827, cast iron was first used to build both a lock and its gates in Cheshire, England. Following the development of the Bessemer process for mass production of steel, that material was utilized for construction of locks and gates, both as primary elements and as reinforcing bars for concrete.

Raw Materials

Waterproof linings keep a canal's water from seeping into the ground. For many years, the best choice was puddle, a mixture of sand, clay, and water that dried to a waterproof state. Modern materials and additives that are more durable include concrete, fly ash, bentonite, bituminous materials, and plastic sheeting.

Locks are usually made of concrete, occasionally lined with steel. If construction of the lock exposes bedrock, the floor need not be lined. The gates are made by welding together steel plates and reinforcement beams. The vertical edges of the gates are fitted with effective sealing materials such as white oak. In 1999, a French company developed lock gates made of glass-fiber reinforced plastic laminate mounted in stainless steel frames.

Design

Early canals followed the most level surface route possible because large-scale earth-moving was so difficult and expensive. Better excavation equipment and lock construction capabilities permit the construction of shorter, more direct canal routes. Because of geographic obstacles, sections of some canals are built in tunnels or on aqueducts (water-carrying bridges).

It is important to fill and empty a lock chamber while producing minimal water turbulence. Modern designs place sluices in the gate sills or in the chamber's floor or walls. A lock may also be equipped with a submerged bubbler that releases air below the gate closure area; the resulting gentle turbulence keeps the area clear of debris that might prevent the gate from sealing properly.

Various gate designs are available for modern locks, and different types may be used on the upstream and downstream ends of a single lock. Miter gates are one of the most popular choices. Another common choice is the Tainter gate, a curved plate that rotates vertically. In this efficient design, which is used for valves in water-transfer culverts as well as main lock gates, water pressure actually assists in the gate's rotation. Flat gates that slide up, sideways, or down into the lock floor are other options, as are flat gates that are hinged at the bottom and curved gates that rotate horizontally into wall recesses.

William C. Gorgas was born October 3, 1854, near Mobile, Alabama. In 1875 Gorgas received a bachelor of arts degree from the University of the South. Desiring a military career he decided to enter the Army with a medical degree. After graduating from Bellevue Medical College in New York and interning at Bellevue Hospital, Gorgas was appointed to the Medical Corps of the U.S. Army in June 1880. Tours of duty followed in Texas and North Dakota, with nearly 10 years at Fort Barrancas in Florida—Gorgas was assigned this yellow fever area because he had previously had the disease and was immune.

After Havana, Cuba, was occupied by American troops in 1898, Gorgas took charge of a yellow fever camp at Siboney, soon becoming chief sanitary officer of Havana. Acting on information that a certain mosquito carried yellow fever, Gorgas quickly destroyed that mosquito's breeding ground, ridding the city of yellow fever.

In 1904, work commenced on the Panama Canal. Gorgas went to the Canal Zone to take charge of sanitation, succeeding in Panama and Colon. Gorgas came to be generally regarded as the world's foremost sanitary expert, and a number of foreign governments and international commissions sought his aid. His book Sanitation in Panama quickly became a classic in the public health field. In 1914 he was appointed surgeon general of the Army, serving in that capacity until his retirement four years later. Gorgas died on July 3, 1920, and is buried in Arlington National Cemetery.

THE MANUFACTURING PROCESS

Canal

- 1 A thorough survey of the canal route must be made, not only for correct alignment but also for accurate depths of cuts and fills.
- 2 Primary excavation is done with earth-moving equipment like bulldozers and excavators.
- 3 A crawler-track mounted trimming machine is used to remove the final 12-18 in (30-46 cm) of soil to create the desired wall slopes and flat bed for the canal. Excavated soil is carried by conveyor and loaded onto trucks that will haul it away.
- 4 A layer of highly permeable material may be spread over the walls and floor to promote drainage of groundwater under the canal.
- 5 Grids of reinforcing steel are built and llifted into place on the canal walls and floor. Concrete blocks hold the grids above the surface so concrete can flow under them.
- 6 Concrete is mixed, vibrated to remove air pockets, and applied with a slipform machine, encasing the reinforcing steel. Slipforming means pouring concrete between the canal surface and a form (mold) that slowly moves forward; the concrete hardens quickly enough to retain

its shape by the time the form moves out of contact with it. Specialized machines place transverse and longitudinal expansion joints in advance of the slipform machine.

Lock

- 7 A temporary cofferdam is built around the proposed lock site. Steel sheet piles are driven into the ground to form a series of adjacent, vertical cells that extend above the waterline. The cells are filled with sand. Water is pumped out of the enclosed space to create a dry construction area. The lock site is excavated. If necessary, piles are driven into the ground to support the lock structure.
- 8 Wooden forms are built to shape the floor and walls of the lock. Space for culverts and valve chambers is included within the forms, as are slots for the gate hinges and recesses to contain open gates. Additional forms are built to shape the approach walls that will guide vessels into the lock.
- 9 After reinforcing steel cages are constructed in the forms, concrete is poured. When the concrete has cured (hardened), the forms are removed.
- 10 Control valves for filling and emptying the lock are installed, along with hydraulic and mechanical equipment for operating the valves and gates.
- 11 Gates are prefabricated and shipped to the site. Very large gates may be shipped in sections, which are welded together as they are installed in the lock.
- 12 Accessories such as guardrails, mooring posts, and escape ladders are installed on the lock walls.

Byproducts/Waste

Disposing of the excavated material is one of the challenges of canal and lock construction. It may be used to construct embankments or be spread over the surrounding countryside where it is carefully landscaped for erosion control as well as appearance. Construction of the Divide Cut, a 29-mi (46-km) canal on the Tennessee-Tombigbee Waterway, in the early 1980s required the disposal of 150 million cubic yards (115 million cubic meters) of dirt.

During lock construction, excavated material can be used to fill cofferdam cells. When the cofferdam is removed, the material may be used to fill in behind the riverbank side of the lock wall. Lock operation uses large amounts of water; for example, it takes 3.5 million gal (13 million L) of water to fill a 600-ft (180-m) long, 110-ft (34-m) wide lock with a 7-ft (2.1-nm) lift. On some canals and rivers, water supplies are limited and conservation is important. Rather than being released downstream as the lock is emptied, some of the water may be diverted to a side pond, where it can be stored and used to help fill the chamber for its next operation.

The Future

Because of the significant need to recondition or replace locks as well as build new facilities, the U.S. Army Corps of Engineers sponsors and conducts research on water-way construction. Scale models of proposed designs can be tested at its Waterways Experiment Station. Through its Innovations for Navigation Projects (INP) Research Program, it supports the development of new ways to build, repair, and operate canals and locks. Recent topics included improvements in underwater concrete and grout placement and development of low-density, high-strength concrete for building modular lock sections that can be installed without construction of a cofferdam. Other current research topics involve protection devices for lock walls and gates from barge impacts, improved designs for water intake and discharge systems, better equipment and control devices for operating gates, and techniques for enlarging (rather than replacing) existing locks that have inadequate capacity.

CANAL - CONSTRUCTION AND OPERATION

Engineers who are designing a canal must take several things into consideration. They must formulate the dimensions of the canal to accommodate the numbers and sizes of ships that are predicted to use the waterway. Natural obstacles in the path of the canal, such as rock formations, must also be modified, removed, or avoided. There must be adequate vertical clearance above the canal and the clearance afforded by pre-existing bridges must be able to accommodate the vessels that will use the canal. Finally, engineers must decide the scale and location of associated structures, such as bridges, tunnels, and locks.

The paths of most canals are affected by variations in the levels of terrain. Engineers compensate for these variations with either locks or inclined planes. A lock is a segment of the waterway that is closed off by gates at either end. When a vessel enters the lock, the front gate is already closed. The back gate is then closed behind the vessel, and the water level within the lock is raised or lowered to the level of the water on the outside of the front gate. Valves on the gates control the level of the water. Whereas locks are the most common means of compensating for elevation changes, the procedure is slow and uses a large amount of water.

Inclined planes can be used to elevate and lower smaller vessels; they use no water and often allow more rapid passage of vessels. When vessels reach certain stations along the waterway, they are pulled out of the water and moved on trucks up or down the plane.

Canal operators must monitor the canal's supply of water. If the natural supply of water at the upper end of the canal is deficient, it must be supplemented by water pumped into the reservoirs. If nature supplies the reservoir with excess amounts of water, some of the water must be diverted from the canal. Otherwise, excess water may strengthen the current and disrupt canal operations.

Although canals are among the oldest civil works, they play a major role in commerce because they are by far the least expensive form of inland transportation yet devised.

Steps to complete Canal Construction

Identity cards had been issued to 2.02 lakh agriculture labourers under Tamil Nadu Agriculture Labourers Social Security scheme in the district so far, said the Collector, Har Sahay Meena.

Distributing identity cards to farm labourers at a function held at Theni-Allinagaram Cooperative Market here on Friday, he said agri-labourers and tiny and small farmers aged between 18 and 65 years had been included in the scheme.

Small farmers having 2.5 acres of wetland and five acres of dryland were eligible for this scheme.

The eligible beneficiaries need not pay any amount for joining this scheme. Each beneficiary, in case of death, would get Rs.1 lakh as compensation. A sum of Rs.50,000 would be given as compensation to the beneficiaries who had lost their limbs or eyes in accidents. The injured would get Rs.25,000 and family member of the beneficiary would receive Rs.10,000 in case of natural death of the beneficiary.

Marriage assistance of Rs.2,000 would be given to male member and Rs.5,000 to female member. Maternity assistance of Rs.2,500 would be given to women members and educational assistance ranging from Rs.1000 to Rs.6,500 to the wards of members.

All efforts would be made to complete the construction of 18th canal as early as possible, he said, adding that the work at Cumbum and Gudalur had been completed.

"We have been waiting for approval from the Forest Department for acquisition of land to construct a dam across the Vairavanar river."

Mr. Meena ordered the officials to expedite acquisition of land for constructing a canal at Melagudalur and Uthamapalayam areas.

On completion of the work, 44 old irrigation tanks and four new tanks would receive water and 4,704 acres would be irrigated.

This canal would pass through Gudalur, Cumbum, Uthamapuram Pudhupatti, Kombai and Thevaram, mostly dry areas, and connect with the Sutha Ganga stream.

Later, the Collector disbursed accident benefit scheme to four persons and identity cards to new members.

Design of Lined Canals

The lined canals are not designed making use of Lacey or Kennedy Theory because the section is rigid. Generally Manning's equation is used in design. To carry a certain discharge number of channel sections may be designed with different bed widths and side slopes. But it is clear that each section is not equally good for the purpose.

The section to be adopted should be economical and at the same time it should be functionally efficient. It has been found that the most suitable cross-section of a lined canal is a circular section with sloping sides. That is, the bed is not flat but it is an arc of a circle. This arc is tangential to the sloping sides.

Side Slopes:

The side slope is selected in such a way that it nearly equals the angle of repose of the soil in the subgrade. Care is taken to ensure that no earth pressure is exerted on the back of the lining. From the knowledge of hydraulics it is clear that the section is economical when cross- sectional area is maximum for minimum wetted perimeter.

This condition is achieved when the centre of an arc lies at FSL of the canal. This section is also efficient in the sense that as the velocity of flow is higher silt carrying is also higher than a wide and shallow section. Thus the problem of silting is completely eliminated and functioning is efficient.

It may be mentioned here that such section with circular bed may be designed up to a discharge of 85 m/sec. When the discharge is more than 85 m/sec the section best suited is one with a flat bed and sloping sides with rounded corners. This section is certainly better than trapezoidal section, because it is more stable and economical to construct. Dimensions of a canal section with circular bed may be obtained from the equations given below. Figure 10.2 shows the canal sections of this type with two standard side slopes.

When r = 3.6 m or less, side slopes may be taken 1: 1

When r = 3.6 m side slopes may be taken 1.25: 1

Velocity of Flow:

Mean velocity of flow may be calculated using Manning's formula.

V = 1/N .R. S

where N is coefficient of rugosity and may be taken as 0.018

S is the slope of bed and expressed in fall of bed in m in 10,000 m length

For the given value of N formula may be reduced to

V = 0.556 R. S

where V is velocity of flow in m/sec;

R is hydraulic mean radius in m; and

S is bed slope expressed in metres/metre length,

since (10,000) is merged in the constant 0.556.

Following Table 10.1 gives the sectional data for the two side slopes:

Dimensions of a canal section with flat bottom may be obtained from the equations given below. Figure 10.3 shows the cross-sections of this type with two standard side slopes. It may be mentioned here that generally maximum velocity of flow is recommended to be 2 m/sec and Lacey's 'f is limited to 1.2. When r exceeds 55 m or when discharge is more than 85 m/sec, the canal should be designed with a flat bottom with fillets at corners. The Indian Standard 4745 in fact does not suggest lined canal with circular bottom.

Velocity of Flow

Mean velocity of flow may be calculated from Manning's formula for value of N = 0.018.

Then V = 0.556 R. S

Following Table 10.2 gives sectional data for the two side slopes:

It may be remembered that lined canal could be given steep slope to achieve recommended velocity of 2 m/sec and value of 'f' about 1.2. Critical velocity ratio is not applicable to lined canals. But to avoid possibility of sitting CVR should be aimed at more than unity.

Coefficient of Rugosity (N)

In general practice for lined canal average value of N may be taken as 0.018. For different types of linings the value of W varies. The values given for straight channels in Indian Standard 4745 are given in Table 10.3. When the alignment is not straight loss of head increases and a small increase in the value of W may be made to allow for additional loss of energy.

Freeboard

Freeboard is measured from the (FSL) full supply level to the top of lining. For lined canals having less than 10 cumec discharge 0.6 m free board is recommended. For bigger lined canals freeboard not less than 0.75 m is generally provided.

Bank Widths

The Indian Standard recommended following values for bank widths for main and branch canals:

Main canal in cutting and filling = 8.0 m

Branch canal in cutting = 6.5 m

Branch canal in filling, left bank = 6.5 m

Right bank = 5.0 m

Suitable provision of dowel, roadway, catch water drain, under drainage has to be made for every lined canal. Figure 10.4 shows three typical cross-sections of the lined canal in which provision of various components is illustrated.

The canal sections shown have trapezoidal shape. For circular bottomed canal arrangement will be similar. It may be noted that angle 0 shown in the section varies with the side slopes adopted. 9 for side slope 1: 1 has 45° value and for 1.25: 1 is 38° 402 as shown in Fig. 10.3. Maximum height of the spoil bank is limited to 6 m when the excavation is done by machinery. In case work is accomplished manually maximum height is to be restricted to 4 m only. Also when a canal section in filling involves more than 10.5 m filling each section shall be designed and tested for stability.

9

INLAND WATERWAYS

Inland waterway transport is generally acknowledged as a clean, safe, and most energy efficient mode of transport when compared to land-based systems which are often confronted with congestion and capacity problems.

Despite the presence of rivers and canals in Asia and the Pacific, until now this potential has remained largely underdeveloped because of lack of dedicated transport policies and investments.

For countries that are developing fast, transport needs are growing even faster. This problem, according to Menno Menist, managing director of Netherlands-based research and consulting firm Panteia, can be solved through the efficient use of infrastructure.

"Road and rail cannot deal with that in many countries. If you have the opportunity to use inland waterway transport, it would be a very smart move to do so," Menist said. "It's often very environmentally friendly, cost-wise it could definitely be cheaper, and you need every alternative you have to transport goods and passengers."

In India, for example, waterways did not receive the infrastructure funding needed as the government focused on roadways and railways after it gained independence from the British in 1947, explained Jayashree Mukherjee, Vice Chairperson, Inland Waterways Authority of India. "But now, road congestion and traffic constraints on the railways and growing environmental awareness have made inland waterways a very good option."

In the People's Republic of China (PRC), inland water transport has received a growing level of recognition of its importance.

Two major categories under which water transport can be divided are as follows: 1. Inland Waterways 2. Ocean Transport!

Since, prehistoric times, water transport has been used for carrying both men and goods. Water transport probably developed before the use of animals because waterway formed an easy means of travel in places where dense forests on land hindered movement.

The range and importance of water transport was increased when power of wind was harnessed by use of boats or some other medium. At first, boats were small and confined mainly to inland waters and sheltered coastal areas.

The gradual increase in size and complexity of sailing craft allowed trade to be established. The Phonecians, Egyptians, Greeks and Romans as well as Arabs and Indians had wide trading contacts. The use of steam has given a new dimension, greater power and speed to water transport to carry larger goods at long distances.

The use of diesel and other forms of power has changed the entire scenario of the water transport and today most of the world trade is water-borne. The two greatest advantages of water transport are that it uses oceans, rivers, seas and needs no special tracks, and that it is the cheapest form of transport for large and bulky loads.

Water transport can be divided into following two categories:

(i) Inland waterways, and

(ii) Ocean transport.

1. Inland Waterways:

There are three types of inland waterways, namely, rivers, rivers which have been modified or canalised, and specially constructed canals. In earlier times much, perhaps most, of the inland carriage of commodities was by water.

This was possible when vessels were small, the volume of traffic limited, and the time factor was not particularly pressing. But during the 18th century, ships began to grow in size, trade began greatly to expand, and speed of carriage came to be of greater importance.

In order to overcome the limitations of many rivers and, also, to provide many inland towns with water communications, canals began to be built. In England, a pioneer in canal construction, the building of these new man-made waterways became almost a mania. Canal building on the Continent came somewhat later, although there are some examples of early canals, e.g., the Canal du Midi in southern France was constructed in 1681.

In 19th century, there was some decline in inland water transport due to development of railways and road transport. But after sometime, water transport has tended to make a comeback, this is closely linked with the cheapness and capacity for bulk carriage of water transport – witness the recent and current developments of the Albert Canal, the canal between Zeebrugge and Ghent, the canalisation of the Moselle, the Rhone Valley Scheme, etc. Inland waterways have both advantages and disadvantages.

The chief advantages are:

(i) There is no track to lay or maintain, although dredging may be necessary in the case of natural waterways;

(ii) They may provide the only practicable routes, e.g., in very difficult, mountainous country or in areas of very dense tropical forest; and

(iii) Waterways, under favourable conditions, provide cheap transport for heavy, bulky, imperishable commodities such as coal, ore, timber, cement.

The principal disadvantages of inland waterways are:

(i) Rivers may involve devious journeys and may flow in the wrong direction from the point of view of trade;

(ii) Otherwise navigable rivers may be interrupted by falls or rapids while canals require locks if there are differences in level;

(iii) River levels may change seasonally and freezing may occur in winter causing stoppages in navigation;

(iv) Canal construction involves heavy capital outlay and canals require constant maintenance and sometimes dredging, and may also require a water supply;

(v) Transport by water is slow in comparison with most other forms of overland transport and carriage by water is generally unsuitable for perishable produce; and

(vi) Waterways are less flexible than either roads or railways which can more easily adapt themselves to changing industrial location.

Although water transport is carried on to a greater or lesser degree the world over, there are only six major navigable systems of inland waterways: the rivers of Western and Central Europe, the Volga-Don system, the North

American rivers, the Amazon system, the Parana-Paraguay system, and the Chinese waterways.

Inland waterways are best developed in Europe and North America; in other continents their development is moderate.

A brief review of inland waterways is as follows:

Europe

In Europe, France, Germany, Belgium, Netherlands as well as Russia, have very extensive inland waterways including rivers and canals. France is having 5,600 km of navigable rivers and another 4,800 km of canals.

The major French rivers, e.g., Loire, Garonne, Seine, Rhone, Meuse and Moselle have been modified and are linked by canal systems so as to travel entirely by river and canals from Mediterranean Sea to English Channel or from Rhine to Atlantic Ocean. On the other hand, Germany is having 7,040 km of inland waterways.

The important inland waterways of Europe are:

The Rhine Waterway

The Rhine is the busiest navigable river of the world. On both sides of it have developed heavy industries which benefit from cheap water transport. The river is navigable by ocean-going steamers of small size. From the point of economic geography, the rivers prolong the great ocean routes into the interior of lands.

Of the "flowing roads" the Rhine is one of the most remarkable. It leads from the ocean into the very heart of the Continent. The Rhine is a "coal river". The lignite of the Aachen basin and the coal of the Ruhr furnish the greater part of the river freight. The Rhine is one of the rivers most favoured by nature for navigation. The greatest difference between the Rhine from Basle to Strasbourg and the Rhine from below Strasbourg is the heavy gradient in the upstream stretch, which causes a very swift current.

Above Strasbourg the traffic is slight owing to a rapid current, low water and a rock ledge near Istein. But below that point there is a slow current except in the gorge of the Slate Mountain (between Bingen and Bonn). The volume of water is uniform during summer. Rhine navigation is stopped during the months of winter owing to low water.

The Rhine from Basle to Strasbourg is characterised by a torrential regimen betraying its Alpine origin. The seasonal variations – summer floods and winter low water – are the more pronounced and rapid as the gradient increases from

Basle to Strasbourg. A barge which carries a full cargo to Strasbourg in summer must discharge half en route in winter. The Rhine navigators invariably allow a margin of 30 cm between boat-keel and river-bed to guard against sudden fall in the waters which may be very rapid in case of drought.

The Rhine empties into a tributary sea of the North Atlantic which lies at the beginning of the largest current of world commerce. This is a privileged position as compared with most other rivers of Europe. To take advantage of this, other rivers near the Rhine have been connected to it by means of canals. The result is that there is no other river in Europe which has so many canal connections as the Rhine.

Waterways of the Germanic-Baltic Lowlands

An extensive network of waterways consisting of east-west canals joining the north-south flowing rivers crosses the northern German plain. The Mittelland Canal, also known as the Midland Canal was built in 1938, joins the three major rivers of Ems, Weser and Elbe. Kiel Canal is 96 km long, links the Elbe estuary to the Baltic Sea. The Dortmund-Ems canal runs north-south and links the Rhine with ports of Bremen and Emden.

Waterways of Southern Germany

Danube is the main river, which flows through seven countries – Germany, Austria, Czech Republic, Hungary, Yugoslavia, Romania and Bulgaria and navigable for about 2,400 km. Canals like Ludwing Canal, Rhone—Rhine Canal provide a good inland waterway.

In Belgium, total length of inland waterway is 1,535 km. Albert Canal (built in 1940) and also other canals on coastal plain serve the towns of Ghent, Bruges, Zeebrugge and Ostend. Netherlands, at the mouth of Rhine, is criss-crossed by its distributaries, and also has extensive man-made waterways.

The former Soviet Union has developed a system of navigable waterways totaling 1, 44,000 km. Most of these waterways are located in European Russia. The Baltic and White Sea Canal, the Moscow-Volga Canal and Volga-Don shipping canals are nodal.

There are several rivers like Volga, Dvina, Don, Dnieper and Dneister, which are navigable in many parts. But many Russian rivers remained blocked during winter season. In spite of such defects Russian rivers are very important for domestic and foreign trade.

North America

In North America the rivers of most use for navigation are the Mississippi and Missouri and the most important canals are those of St. Lawrence, which unites the Ontario and St. Lawrence; the Sault Sainte Marie canal, between Superior and Huron; the canal which links the Chesapeake to the Ohio; the New York canal; and the canals between North Allegheny and Erie.

The length of navigable waterways in United States is over 36,072 km. The Mississippi river system, largest of all, provides more than 8,000 kilometres of waterways with depths of 3 metres or more, including the main river trunkline from Minneapolis to the Gulf of Mexico — a distance of more than 12,880 kilometres. The Missouri, a tributary of Mississippi, is navigable for 1,216 km to Sioux City, Iowa.

On the recommendations of The Ralph M. Parsons Company, a private engineering and construction enterprise with headquarters in Los Angeles was set up under the project to carry out technical assessment of water and power potential of North America. The project is referred to by the company as NAWAPA – the North American Water and Power Alliance.

The basic idea behind 'NAWAPA' is to capture the surplus waters of the Fraser, Yukon, Peace, and Athabaska river systems in north-western North America and to direct, via an elaborate system of canals, reservoirs and tunnels, the surplus water to deficit areas of Canada and United States.

The St. Lawrence Waterway

The St. Lawrence, with the great lakes, forms a very important commercial route into the heart of North America. The river is, however, ice-bound for about four months every year, and has various rapids and falls which have necessitated the cutting of canals to enable sea-going vessels to reach Lake Superior.

Large ocean vessels can pass about a thousand kilometres up the river to Montreal; but here goods have to be transhipped to smaller vessels, because rapids occur, and the canals made to avoid them are not over 3-5 metres deep.

The Canadian government built a 3-5 metres deep canal around the rapids that permitted shallow draft boats to negotiate the St. Lawrence between Lake Ontario and the sea. After the Welland canal and its eight locks were completed in 1931, interest in the Great Lakes — St. Lawrence Seaway was renewed.

The St. Lawrence enters the ocean by a deep estuary due to submergence in the past, but navigation is rendered difficult by the prevalence of fogs and the rapidity of the current. The valley of the St. Lawrence is fertile, and the whole length is lined with villages and towns.

Another canal has been made to avoid the falls of Niagara, though a great deal of trade is diverted at Buffalo to the Erie Canal and Mohawk-Hudson route to New York. The Sault Sainte Marie or 500 canals were necessitated by rapids between Lake Superior and Lake Hudson, and the traffic on these canals is enormous.

Asia

Asia is not having elaborate system of inland waterways, but rivers in many countries are used as inland waterways.

In China, the rivers have made a significant contribution to the development of commerce. The three great rivers, the Hwang-ho, the Yang-tse-kiang and the Sikiang, cross the country from west to east. China's greatest river is the

Yang-tse-kiang, the most important waterway for navigation in the country. It is doubtful whether there is another equally extensive region of wealth in the world where the people depend as solely upon a single artery of traffic and upon one entrepot as do the inhabitants of the Yangtse basin.

About half of the populations of China live in this fertile area, utilising the river, its tributaries and its network of canals as their chief means of communication.

The Yang-tse-kiang rises in Tibet, and with its tributaries drains the heart of China. It is navigable by steamers up to the port of Hankow. The Sikiang rises in the highlands of Yunan am' has a fairly direct course eastward to its mouth. It is navigable for the greater part of its course. The Pei-ho is important for communication and can be navigated up to Tientsin

Northern India is especially endowed with three large navigable rivers. These rivers are the Ganga, the Brahmaputra and the Jamuna. The Ganga can be navigated by steamers as far as Kanpur from its mouth. This river flows through the most densely populated and fertile plain of India and naturally commands much traffic.

Before the development of railways the Ganga was of considerable importance for the movement of goods and persons. The development of railways has greatly reduced the importance of steam navigation, especially in the upper Ganga.

The Lower Ganga is even now very important, and there is traffic all the year round. The Brahmaputra flows through Assam and Bangladesh and is navigable as far as Dibrugarh. Its tributary, the Surma, has made steam navigation possible in Sylhet and Cachar. The Indus in Pakistan is navigable by steamer up to Dera Ismail Khan in the North Western Frontier Province. The river mostly handles wheat, cotton and wool. The frequent shifting of its bed and the formation of sand-bars has caused steam navigation in the Indus to be neglected.

Burma is very fortunate in having a large number of navigable rivers. The Irrawaddy, the most important and the largest, is navigable by steamers for more than 800 km from its mouth and country boats can proceed farther.

Africa

In Africa, some rivers are navigable, that too for a very limited length. The Nile is the most important river in north-east Africa, but its great defect is the succession of cataracts. In its upper course the Nile has rapids and falls; in its middle course there are cataracts. It is navigable in the delta and in its lower course.

The rivers of South Africa are of little use for traffic. The Zambesi is navigable for only 350 km, while the Limpopo can be navigated only for a short distance. The Orange is not navigable. In tropical Africa, the Congo provides a magnificent system ot waterways. It rises in the highlands between the lakes Tanganyika and Nyasa.

But at several places navigation is interrupted by rapids and falls. The Uban the chiel tributary of the Congo can be navigated almost up to its head. In West Africa the Niger is easily navigable for 500 miles and in the wet season navigation is continued further. The Gambia is navigable for 260 km from its mouth.

South America

South America is having some long rivers but their use as inland waterways is limited. Amazon River is the longest river of the continent. But till now the Amazon system is of relatively little use, because the region through which the river flows is densely forested, scantily populated, undeveloped and largely unexplored. The

Orinoco which flows through Venezuela is a long waterway. But the most useful in South America is the Parana system which penetrates the heart of Argentina, Paraguay, Uruguay and South Brazil. In the southern side of South America the river Rio Negro drains the sheep-rearing land of Patagonia.

Australia

Australia is deficient in waterways. Her river-system consists of small streams flowing from the highlands to the coast, thus not suitable for navigation purpose. The two most important rivers are Murray and Darling. Darling River remains almost dry during winter and spring seasons. River Murray is partially used as a waterway.

OCEAN TRANSPORT

Ocean transport is the most important water transport, because it has certain advantages over land carriage. The sea offers a ready-made carriageway for ships which, unlike the roadway or railway, requires no maintenance.

Water surfaces are two-dimensional and, although sea-going vessels frequently keep to shipping lanes, ships can travel, within a limited number of constraints, in any direction.

Because of floatability and reduced friction, ocean vessels are capable of carrying far greater loads and far greater weights than can be handled even by the longest railway train, the most powerful lorry and trailer, or the largest aircraft; accordingly, ocean transport is usually the cheapest of all forms of transport.

Again, except for fog and floating ice, and occasionally stormy weather which may hinder progress, ocean-going vessels have fewer physical obstacles to surmount than those which so often handicap overland transport.

Ships have been used for transport right from early times. During Graeco-Roman times a fundamental distinction was made between longships or galleys, used for military purposes, and round-ships for trade. The Vikings had a similar distinction.

During more recent times ships have shown an even greater specialisation and several distinct types of commercial vessels gradually evolved. Today about half a dozen main types of merchant ships are recognised – passenger liners, cargo-liners, bulk-carriers, tramps and coasters, and short-sea traders.

The most spectacular development, however, has been the appearance of bulk-carriers, the most important of which are tankers, the product of the oil age. Recent years have witnessed a tremendous growth in the size of tankers, a number of which now exceed 5, 00,000 tons dead-weight.

Today, more than half of the world's merchant shipping tonnage comprises tankers, a fact indicative of the great importance of oil in the modern world.

Although ships have freedom of movement and are capable of going virtually anywhere on the ocean surface, they tend to keep to certain "lanes". They do so because of: (i) physical conditions, and (ii) economic considerations.

Clearly, ships will go only where there are goods or people to be carried and the most important shipping routes are those linking the most productive and most populous regions. Certain physical conditions also help to determine the routes followed by ships, e.g., the availability of harbours and ports on coasts, weather conditions such as fog and storm occurrence, and oceanographically factors such as sea ice and icebergs, submarine banks, and shallow waters.

The North Atlantic Ocean Route

The North Atlantic Ocean Route has the greatest traffic of all ocean routes. Nearly one-fourth of the tonnage of the world's merchant vessels serves this route. In volume and variety of cargo, this route far exceeds any other.

This route connects the ports of Western Europe with those on the east coast of North America. These two regions are the most populous and highly developed regions in the world.

North America and Western Europe are the world's greatest producers of goods of quantity and diversity. Ports on the western coast of Europe are Glasgow, Liverpool Manchester, Southampton, London, Rotterdam, Bremen, Bordeaux and Lisbon. Ports on the eastern coast of the USA are Quebec, Montreal Halifax, St. John, Boston, New York, Baltimore, Charleston Galveston and New Orleans.

This oceanic route is the busiest trade route of the world. Large quantities of manufactured items: textiles, chemicals, machinery, fertilizers, steel, wine, etc., are exported from these ports across the North Atlantic to the United States and Canada.

The exports of Canada and the USA to Europe are timber, fish, wheat, raw cotton, tobacco, oil, machinery and vehicles, metals, paper and chemicals.

The Suez Canal or Mediterranean Asiatic Route:

This route is second to the North Atlantic in respect of volume of traffic. It commands the markets of eastern Africa, Iran, Arabia, India, the Far East, Australia and New Zealand. In fact, the route passes through the heart of the world and touches more lands and serves more people than any other route. Throughout its many ports of call, it reaches about three-quarters of the total population of the globe.

After crossing the Red Sea, the route follows two directions – one along the eastern coast of Africa to Durban; another to farther east – to India, Australia, etc.

Ports of departure are London, Liverpool, Southampton, Hamburg, Rotterdam, Lisbon, Marseilles, Genoa and Naples. The ports of call are Aden, Mumbai, Kolkata, Rangoon, Penang, Singapore, Manila, Hong Kong, Perth, Adelaide, Melbourne, Sydney, Mombasa, Zanzibar, Mozambique and Durban. This route is used by Asiatic countries to send raw materials and some food products to the western markets and receive in return manufactured articles – the products of the Far East are rice, tea, sugar and silk; those of India are coffee, tea, pig iron, manganese ore, jute goods, indigo, spices, cotton, teak, silk, skins, leather and oil-seeds and those of Middle East are petroleum, coffee, and dried fruits. From Australia meat, timber, wheat, flour, fruit, wool, butter and wine are sent. China, Australia, New Zealand and countries of South and South-East Asia now use this route for both export and import of commodities.

The Cape of Good Hope Route

This route was once the subsidiary alternative to Suez Canal route, but because of its long and circuitous journey, was avoided by most of the shipping companies. During the closure of Suez Canal in 1967 all the ships had no choice but to take this route.

Even after Suez Canal reopened in 1975, much trade continues to follow this route because tankers and other vehicles nowadays are much larger. As the Suez Canal can only accommodate ships of around 20,000 tons capacity and toll charges are high, the Cape route is growing in importance.

It has several other advantages. With the greater economic development of the recently independent African nations and the exploitation of their rich natural resources such as gold, copper, diamonds, tin, chromium, manganese, cotton, oil palm, groundnuts, coffee and fruits, the volume of traffic round the Cape of Good Hope and from ports in both East and West Africa is on the increase.

The Panama Canal: West Indian Central American Route

The construction of Panama Canal was completed in 1913. The Panama Canal is 'the gateway to the Pacific' and eliminated the long and hazardous voyage round the Cape Horn. It has benefited on both Atlantic and Pacific seaboard, facilitating trade in minerals, oil, foodstuffs, raw materials, and manufactured products. But the greatest benefits have accrued to traffic between the east and west coasts of USA.

The Panama route has also greatly facilitated trade in the West Indian islands and the Pacific states of North, Central and South America, especially the Andean states which are rich in mineral resources and have good markets in North America. The Latin American states import manufactured goods and mining equipment from the USA and the western European countries.

Much trade destined for the Far East, the Pacific islands and Australasia from North America and Western Europe also goes through the Panama Canal. With the greater economic development of East Asian countries especially China, Japan and the South-East Asian states, the Panama route is assuming a greater role in the exchange of products between the East and West. The distance saved from Auckland to New York via Panama, instead of Cape Horn, is more than 4,000 km.

The South Atlantic Route

This route leads to West Indies, Brazil and Argentina. The chief ports of call on the route are Kingston (Jamaica), Havana, Vera Cruz, Tampico, Pernambuco, Bahia, Rio de Janeiro, Santos, Montevideo, Buenos Aires and Rosario. The exports along the route are sugar, bananas, raw cotton, mahogany, tobacco, coffee, grain, wool and meat, while the imports are manufactured goods.

This route maintains trade connections between Europe on the one hand and West Indies, Caribbean seaboard, Brazil, Uruguay and Argentina on the other.

The Trans-Pacific Route

There are several routes in North Pacific which converge at Honolulu for refuelling and servicing. The direct route further north is a great circle which links Vancouver and Yokohama without calling the Hawaiian Island, reduces the travelling distance by half.

The North Pacific trade includes Vancouver, Seattle, Portland, San Francisco and Los Angles on the American side, dealing with wheat, timber, paper and pulp, fish, dairy products and manufactured goods. The destinations across the 7,200 km (4,500 mile) wide Pacific are usually Yokohama, Kobe, Shanghai, Guangzhou (Canton), Hong Kong, Manila and Singapore. The east-bound trade from East Asia to North America consists mainly of manufactured goods, e.g., textiles, electrical equipment, from Japan, Hong Kong, S. Korea, and Taiwan, and tropical raw materials from South-East Asia, e.g., rubber, copra, palm oil, teak, tin and tea. In addition to international links the North Pacific is an important domestic route-way from the US mainland to the isolated states of Alaska, in the north, and Hawaii in the mid-Pacific.

In the South Pacific, the traffic consists mainly of ships travelling via the Panama Canal between either West Europe or North America and Australia, New Zealand and the scattered Pacific islands. Goods transported are mostly wheat, meat, wool, fruits, dairy products and manufactured articles. The other important routes are eastern North American-east South American (from New York to Cape Sao Roque), North American-western South American (from New York to Punta Arenas via Panama Canal), North American-Australasian (from New York and Vancouver to Sydney and Wellington via Honolulu), etc.

The analysis of ocean transport cannot be completed without detailed description of Suez and Panama Canals. These two canals have changed the entire pattern of oceanic trade, therefore, it is essential to know the characteristics and importance of these canals.

The Suez Canal

The Suez Canal is one of the great international waterways of the world — cuts across the Isthmus of Suez and provides navigational facilities between the Mediterranean Sea and the Indian Ocean.

The history of a canal connecting the Mediterranean with the Red Sea dates back to 13th century B.C. when Nile-Red Sea canal is known to have been in use until the end of 8th century AD. From 16th century onward one or the other of European powers became interested in the idea of either reopening the old waterway or cutting a new one from the Mediterranean.

In 1834, Ferdinand de Lesseps, a member of the French Consular service at Alexandria was interested in the Suez Canal scheme. In 1854, he discussed the project with the Viceroy of Egypt (Khedive) and got his approval. A concession to run for 99 years from the date of the canal's opening was granted to de Lesseps, authoring him to form an International Company for the purpose of constructing a waterway.

The excavation was started in 1859 under de Lesseps, who took full ten years to construct the canal. It was opened in November, 1869. It is about 160 km long (including distance of lakes) and is 11 to 15 metres deep. The width of the floor is 40 m and varies at the surface. Suez Canal links the Mediterranean Sea with Red Sea. Port Said is located at Mediterranean Sea; while Port Suez is on the Red Sea.

As a ship enters the canal from the Mediterranean Sea, it will pass Port Said, one of the greatest ports in the world and proceeding south will enter Lake Tunisa on the bank of which is the city of Ismailia. From Lake Tunisa to Suez city, a ship will pass through Geat Bitter Lake and Little Bitter Lake.

No single human enterprise during the past century has done more to affect the destinies of nations through a physical geographical change than the Suez Canal. The opening of Suez Canal route saved approximately 5,820 kilometres on the voyage from London to Mumbai as compared to the Cape route. The opening of Suez Canal had a tremendous affect not only on world trade and commerce but also on international politics, besides opening up to western countries a new route to Africa, Asia and Australia.

The usual trade route between the eastern coast of North America and the Far East was through the Cape of Good Hope. The Suez Canal saved a great deal of distance by diverting the traffic from the Cape of Good Hope route to itself and thus, benefited North America greatly. More than 12,000 vessels pass through the Suez Canal every year.

The Suez Canal has provided not only the fastest but also the most economical line of transit between Europe and the East. Politically the Suez route is vital because of the oilfields in the Middle East countries on whose products the economy of Western Europe is dependent.

There are some problems of the Suez Canal. The canal needs improvements in regard to depth, width, diversion to avoid crossing of ships in the narrow part of canal. The deposition of silt that comes along winds blown from desert also needs regular cleaning. The second problem is the high canal dues levied on ships which pass through the canal.

It has been observed that when speed is not essential, many cargo liners follow the Cape of Good Hope route to avoid the high dues. Now many cargo vessels are so big in size that they cannot pass through the Suez Canal. Political instability in adjoining states of the canal is also a cause of concern. Although, Suez Canal according to international convention is free and open, in time of war as in time of peace, to every vessel of commerce or of war, without distinction of flag.

The Panama Canal

The Panama Canal connects the Atlantic and the Pacific oceans by means of two bays, an artificial lake, a natural lake, and three systems of locks. It has been constructed across the narrow Isthmus of Panama where the long Continental Divide dips to one of the lowest points. The canal is 72 km long from deep water to deep water in the oceans. It was opened on 15 August, 1914.

All the locks are double, so that ships can pass in both directions without any congestion of traffic. The depth of the channel varies from 12 to 26 metres and the width varies from 91 to 305 metres. The time taken to pass through the canal from Panama to Colon is 14 hours.

The Panama Canal passes through a rough country and the engineering difficulties have been much greater than in the case of the Suez Canal, which passes through a level country and needs no locks. The Panama Canal generates its own hydroelectric power with which not only lighting of the region is done, but also electric locomotives are supplied to pull ships through the locks.

The Panama Canal is 'the gateway to the Pacific'. It has benefited countries on both Atlantic and Pacific seaboards, facilitating the trade in minerals, oil, foodstuffs, raw materials and manufactured products. Its greatest benefits have accrued to traffic between the east and west coasts of the United States.

The canal shortens the distance between New York and San Francisco by about 12,596 km by sea, and lessens very considerably the distance between Western Europe and western America, and between the northern and middle parts of East America and East Asia.

It also slightly shortens the distance between Europe and New Zealand, but it does not lessen that between Europe and Asia or Europe and Australia. Eastern North America and north-western Europe very definitely gained as a result of the canal, for they were placed much nearer by this route to all of western North America, western South America, and New Zealand.

For eastern North America the canal has meant a great reduction in the distance to Japan and to all of China north of Hong Kong, a factor that has unquestionably contributed to the rapid growth of trade with East Asia.

INLAND WATERWAYS OF THE UNITED STATES

The inland waterways of the United States include more than 25,000 mi (40,000 km) of navigable waters. Much of the commercially important waterways of the United States consist of the Mississippi River System—the Mississippi River and connecting waterways.

The Columbia River is the only river on the West Coast (and arguably the entire North American Pacific coast) that is navigable for a significant length. The river is regularly dredged and freight barges may reach as far inland as Lewiston, Idaho through a system of locks; however, there are strict draft restrictions beyond the confluence with the Willamette River. The Sacramento and San Joaquin Rivers, the Snake River, and the Umpqua River are examples of other West Coast rivers that are dredged for navigation. The steep grades and variable flows of most other West Coast rivers make them unsuitable for large boat travel. Also, most large rivers there are dammed, often in multiple places, to supply water for hydroelectricity production and other uses. Mountainous terrain, and a shortage of water, make canals in the West unfeasible as well.

Most navigable rivers and canals in the United States are in the eastern half of the country, where the terrain is flatter and the climate is wetter. The Mississippi River System, including the Gulf Intracoastal Waterway (GIWW) connects Gulf Coast ports, such as Mobile, New Orleans, Baton Rouge, Houston, and Corpus Christi, with major inland ports, including Memphis, Kansas City, St. Louis, Chicago, St. Paul, Cincinnati, and Pittsburgh. The Lower Mississippi River from Baton Rouge to the Gulf of Mexico allows ocean shipping to connect with the barge traffic, thereby making this segment vital to both the domestic and foreign trade of the United States. The Mississippi River System is connected to the Illinois Waterway, which continues to the Great Lakes Waterway and then to the Saint Lawrence Seaway. Many other eastern rivers are navigable as well, including the Potomac, the Hudson, and the Atchafalaya Rivers, which are all dredged by the Army Corps of Engineers.

Navigable waters included in legal definition

Title 33 of the United States Code and 33 C.F.R. define the "navigable waters of the United States" and apply certain laws and regulations to those waters. This determination is made by a combination of waters explicitly listed in the law, and general definitions that mean certain waters might or might not be included depending on various factual determinations (such as being "navigable in fact" and the history of use) by the Army Corps of Engineers. Not all waters have had these facts determined, and so are of uncertain status. All water subject to tides are included.

Note that the "Navigable Waters of the United States" listed in 33 C.F.R. 329 are different than those listed as "Waters of the United States" in 33 C.F.R. 328, which is the Clean Water Rule. However, all Navigable Waters, plus those considered navigable-in-fact are included in the general "Waters" definition.

New England district

Vermont waters have been surveyed thoroughly, but the other New England states have some waters of indeterminate status. Flowing waters are navigable from the mouth to source, or mouth to specified point, unless otherwise noted.

- Kennebec River to Moosehead Lake
- Penobscot River to the confluence of the East and West Branch at Medway, Maine
- Lake Umbagog
- Merrimack River to Concord, New Hampshire

- Connecticut River to Pittsburg, New Hampshire
- Lake Champlain
- Lake Memphremagog
- Ompompanoosuc River to Mile 3.8
- Waits River to Mile 0.9
- Black River to Mile 25 in Craftsbury, Vermont
- Battenkill River to Mile 50 in Manchester, Vermont
- Lamoille River to Mile 79 in Greensboro, Vermont
- Missisquoi River to Mile 88.5 in Lowell, Vermont
- Otter Creek to Mile 63.8 in Proctor, Vermont
- Winooski River to Marshfield, Vermont
- Moose River from Passumpsic River to the border of Victory, Vermont
- Nulhegan River, including the East Branch, Black Branch, and Yellow Branch
- Paul Stream
- Passumpsic River and its East Branch to East Haven, Vermont
- Wells River to Groton Pond
- White River

Though navigable-in-fact, parts or all of the following have been excluded from the definition by Congress:

- Park River in Hartford County, Connecticut
- Burr Creek in Bridgeport, Connecticut
- Boston Inner Harbor, Fort Point Channel and South Bay in Boston
- Acushnet River in the harbor of New Bedford, Massachusetts and Fairhaven, Massachusetts
- West River in West Haven, Connecticut<ref.33 USC 59</ref>
- Back Cove in Portland, Maine
- Kenduskeag Stream in Penobscot County, Maine

EFFICIENCY

A principal value of the inland waterways is their ability to efficiently convey large volumes of bulk commodities moving long distances. Towboats push barges lashed together to form a "tow". A tow may consist of 4 or 6 barges on smaller waterways and up to over 40 barges on the mighty Mississippi River below its confluence with the Ohio River. A 15-barge tow is common on the larger rivers with locks, such as the Ohio, Upper Mississippi, Illinois and Tennessee rivers. Such tows are an extremely efficient mode of transportation, moving about 22,500 tons of cargo as a single unit. A single 15-barge tow is equivalent to about 225 railroad cars or 870 tractor-trailer trucks. If the cargo transported on the inland waterways each year had to be moved by another mode, it would take an additional 6.3 million rail cars or 25.2 million trucks to carry the load.

The ability to move more cargo per shipment makes barge transport both fuel efficient and environmentally advantageous. On average, a gallon of fuel allows one ton of cargo to be shipped 180–240 mi (290–390 km) by truck (e.g. @ 6–8 mpg US (2.6–3.4 km/l) 30 ton load, 450 mi (720 km) by railway, and 514 mi (827 km) by barge. Carbon dioxide emissions from water transportation were 10 million metric tons less in 1997 than if rail transportation had been used. Inland waterways allow tremendous savings in fuel consumption, reduced greenhouse gas emissions and air pollution, reduced traffic congestion, fewer accidents on railways and highways, and less noise and disruption in cities and towns.

Commodities

Barges are well suited for the movement of large quantities of bulk commodities and raw materials at relatively low cost. The inland and intracoastal waterway system handles about 630 million tons of cargo annually, or about 17 percent of all intercity freight by volume. These are raw materials or primary manufactured products that are typically stored for further processing or consumption, or transshipped for overseas markets.

1. Coal is the largest commodity by volume moving on the inland waterways. The country's electric utility industry depends on the inland waterways for more than 20 percent of the coal they consume to produce electricity.
2. Petroleum is the next largest group, including crude oil, gasoline, Diesel fuel, jet fuel, heavy fuel oils and asphalt.
3. Another large group includes grain and other farm products, most of which moves by waterway to ports on the Lower Mississippi River or

Columbia River for export overseas. Sixty percent of the country's farm exports travel through inland waterways.

4. Other major commodities include aggregates, such as stone, sand and gravel used in construction; chemicals, including fertilizers; metal ores, minerals and products, such as steel; and many other manufacturers products.

Economic value

Inland and intracoastal waterways directly serve 38 states throughout the nation's heartland as well as the states on the Atlantic seaboard, the Gulf Coast and the Pacific Northwest. The shippers and consumers in these states depend on the inland waterways to move about 630 million tons of cargo valued at over $73 billion annually. States on the Gulf Coast and throughout the Midwest and Ohio Valley especially depend on the inland and intracoastal waterways. Texas and Louisiana each ship more than $10 billion worth of cargo annually, while Illinois, Pennsylvania, West Virginia, Kentucky, Mississippi, Alabama, and Washington state each ship between $2 billion and $10 billion annually. Another eight states ship at least $1 billion annually. According to research by the Tennessee Valley Authority, this cargo moves at an average transportation savings of $10.67 per ton over the cost of shipping by alternative modes. This translates into over $7 billion annually in transportation savings to economy of the United States.

Maintenance and modernization

The U.S. Army Corps of Engineers (USACE) is responsible for 12,000 mi (19,000 km) of the waterways. This figure includes the Intracoastal Waterway. Most of the commercially important inland waterways are maintained by the USACE, including 11,000 mi (18,000 km) of fuel taxed waterways. Commercial operators on these designated waterways pay a fuel tax, deposited in the Inland Waterways Trust Fund, which funds half the cost of new construction and major rehabilitation of the inland waterways infrastructure.

The nearly 12,000 miles (19,000 km) of U.S. inland and intracoastal waterways maintained by the U.S. Army Corps of Engineers includes 191 commercially active lock sites with 237 lock chambers. Some locks have more than one chamber, often of different dimensions. These locks provide the essential infrastructure that allows tows to "stair-step" their way through the system and reach distant inland ports such as Minneapolis, Chicago, and Pittsburgh. The locks can generally be categorized by three different sizes, as expressed by length. About 15 percent of the lock chambers are 1,000 to 1,200 ft (300 to 370 m) long, 60 percent are

600 to 999 ft (183 to 304 m) long, and 25 percent are less than 600 feet (180 m) long. Lock widths are mostly 110 feet (34 m). The 1,200-foot (370 m) locks can accommodate a tow of 17 barges plus the towboat, while the 600-foot (180 m) locks can accommodate at most eight barges plus the towboat. The lock size and tow size are critical factors in the amount of cargo that can pass through a lock in a given period of time.

More than 50 percent of the locks and dams operated by the U.S. Army Corps of Engineers are over 50 years old. Many of the 600-foot (180 m) locks on the system were built in the 1930s or earlier, including those on the Ohio, Upper Mississippi, Illinois and Tennessee rivers. These projects are approaching the end of their design lives and are in need of modernization or major rehabilitation. Since many of today's tows operate with 12 or more barges, passing through a 600-foot (180 m) lock requires the tow to be "cut" into two sections to pass the lock. Such multiple cuts can be time consuming and cause long queues of tows waiting for their turn to move through the lock.

In the 1960s the U.S. Army Corps of Engineers began to modernize the locks on the Ohio River and added 1,200-foot (370 m) chambers that permit a typical tow to pass in a single lockage. This modernization process continues today with the construction of a new dam with twin 1,200-foot (370 m) locks at Olmsted, Illinois located at the confluence of the Ohio and Mississippi rivers and a second 1,200-foot (370 m) chamber at McAlpine Locks and Dam near Louisville. Modern 1,200-foot (370 m) chambers are also being constructed at Kentucky Lock on the Tennessee River and at the Inner Harbor Lock on the Gulf Intracoastal Waterway at New Orleans. Other projects are underway in Pennsylvania, West Virginiaand Arkansas. In addition, several major rehabilitations are also underway. Altogether, this ongoing work represents an investment of over $3.5 billion in inland waterway modernization that will be completed over the next decade. Half this investment will come from fuel taxes paid by the inland towing industry. These projects include not only modern navigation facilities, but also important investments in environmental restoration and management.

Several key navigation improvement feasibility studies are underway throughout the inland waterway system, including on the Upper Mississippi River and Illinois Waterway, Ohio River, the Gulf Intracoastal Waterway, the Black Warrior River and the Tennessee River. Over the next few years, these studies will identify the navigation and natural environmental actions needed to support the inland waterway system. While annual capital spending for the inland waterway system has averaged about $170 million in recent years, the income stream from fuel tax revenues can support an annual capital investment program of about $250 million without reducing the surplus in the Inland Waterways Trust Fund, whose balance was $385 million at the end of 1999.

A TRANS-AFRICA INLAND WATERWAY SYSTEM

The Nile River has been navigable since the time of the pharaohs some 4,000 years ago, as have sections of the Niger, Benue, Congo and Zambezi Rivers.

Boats today carry passengers and freight across Lake Malawi, Lake Victoria and Lake Tanzania while riverboats provide essential services along sections of the Niger, Benue, Congo and Nile Rivers.

There may be scope for future African leaders to borrow the European precedent of connecting navigable rivers to develop an inland, canal-based transportation network.

Inland waterway systems have also been successful outside Europe. In the U.S., inland waterway freight transportation can move bulk and container shipments of over 100 TEUs at lower cost per unit distance than either truck or railway transportation.

Such a system would benefit the African continent, which remains largely dependent on fragmented land transport systems. While Cecil John Rhodes sought to build a North – South, Cape-to-Cairo railway line, African railway networks remain regional with their own distinctive railway gauge. Paved road networks are to be found in major cities and in certain regions, but only a network of unpaved roads is available for most of the continent's long distance connections for the truck transport industry. Trucks carry most of Africa's long-distance freight.

The headwaters of several streams and rivers that empty into Lakes Tanzania and Victoria originate in the same geographic area between the lakes, in sufficiently close proximity to each other to perhaps warrant linking them by building navigable canals to transit shallow draft vessels.

Lake Victoria empties through a series of smaller lakes and waterfalls into the White Nile River, perhaps with the potential to build navigation locks to connect the Upper Nile River to Lake Victoria. Lake Tanzania empties into the headwaters of the Congo River that has several navigable sections. There may be scope to develop river navigation between Lake Tanzania and the navigable sections of the Congo River, allowing access to Lake Victoria. It may also be possible to develop future canal navigation between navigable sections of the Congo River and its mouth at the Atlantic Ocean.

At the present time, Nigeria and Egypt have surpassed South Africa as Africa's leading economies. A navigable inland waterway system could offer a much shorter sailing distance than an ocean voyage through the Strait of Gibraltar. In southern Sudan and northeastern Congo, headwaters flow from the same region into the White Nile and Congo Rivers, allowing for possible evaluation of building a navigable canal to link the two rivers.

The development of a trans-Africa navigable inland waterway system would require political agreement and political cooperation between several national governments. While some regional governments operate and subsidize the operation of railways, they collect a tariff from the truck transportation sector and may be expected to do likewise on vessels that sail along waterways and through navigation locks located within their respective jurisdictions. Vessel operators would likely have to pay transit fees to pass through navigation locks along the future navigable waterway. It may be possible for inland water transportation to be competitive with trucks to transport freight.

During the time of the pharaohs, navigable canals were believed to connect the Nile River to the Red Sea. A redevelopment of such a waterway would connect a possible inland trans-Africa waterway to ports along the coast of Saudi Arabia and as far north as the Gulf of Aqaba.

The dredging precedent of the Suez Canal may be possible for some distance between the Red Sea and Nile River. Modern water-pumping technology can move massive volumes of water to higher elevations and facilitate canal navigation from the Red Sea close to the Nile River.

In the southeastern region, a section of the Zambezi River is navigable with the potential to develop canal navigation north to Lake Malawi. There may be a basis to construct a navigable canal between Lakes Malawi and Tanzania, using the beds of several rivers and streams that flow in the area between the lakes, in northeastern Zambia. It may further be possible to develop navigation between two large dams on the Zambezi River, Kariba and Cabora Bassa. Construction of special canals along a series of rapids could provide river navigation to the Indian Ocean.

To achieve a major revamp of Africa's transportation systems, engineering students attending colleges and universities across Africa could conduct research into the possibility of a modern trans-Africa inland waterway transportation network.

Inland Water Transport and Ocean or Sea Transport available in India

Water transport of India is divided into two types. These are Inland Water Transport and Ocean or Sea Transport.

Inland Water Transport

Inland waterways include transport through navigable rivers and canals. Different rivers of India possess 5000 kilometers of navigable waterways but we use only 1700 km Of it. We use boats and steamers in land waterways. Canal water-ways are not in a satisfactory condition. Out of 4300 km of canal ways 485 km are suitable for motor boat and out of that 331 kms are actually used for motor transport.

Among the different navigable rivers of India, the Ganga, the Brahmaputra and their sub-rivers, the Krishna, the Mahanadi, the Narmada and the Tapti are important. Bumkimham Canal of Andhra Pradesh and Tamilnadu and some canals of Kerala have occupied prominent place in canal ways. States like Assam, West Bengal, Bihar, Uttar Pradesh, Andhra Pradesh, Kerala, Tamilnadu and Orissa use their inland waterways in a more intensive manner.

Sea or Ocean Transport

The length of coastal area of India is nearly 6100 km Navigation along the coastal places and foreign countries take place with the help of ships. 55 shipping companies are operating for ship transport. Shipping Corporation of India is the only nationalised Company and it helps both in external trade and coastal trading.

Sea transport takes place with help of 11 important ports and 165 other small and medium ports. Different important ports are Calcutta, Haldia, Paradeep, Visakhapatnam and Tutikorin in east and Kandala, Mumbai, Marmagoa, Mangalore and Cochin in the west.

External trade of India takes place mainly through four important sea routes. India establishes trade relationship with the countries of Europe "continent through Suez route. India establishes trade relationship with western African countries and South American countries through Uttamasha Cape route. India establishes trade relationship with South East Asia, Japan, the U.S.A. and Canada through Singapore sea route and establishes trade relationship with Australia route.

10

INLAND WATERWAYS OF INDIA

An efficient transport sector is vital for development of the economy of any country. In a large country like India, efficient transportation becomes pivotal to stimulate competitive business environment. Indian transport system comprises various modes, viz; Railways, Roadways, Inland Waterways, Coastal Shipping and Airways.

Inland Water Transport (IWT) is a fuel efficient, environment friendly and cost effective mode of transport having potential to supplement the over burdened rail and congested roads. For this, however, it is necessary that IWT mode is developed with public funding at least to a threshold level at which private sector would get attracted to this mode.

In the 19th century and first half of 20 century, IWT was an important mode of transport and navigation by power crafts/country boats played significant role in the development of trade and commerce along several rivers and canals. The advent of railways and extension of its network made a dent in share of water transport in India. Rapid growth of roads, coupled with inadequate development of IWT sector over the years gave a decisive set back to IWT and in the later years of 20 century, except in a few areas namely, Assam, Goa, Kerala, Mumbai,West Bengal, and some other coastal areas (where it has natural advantage and no developmental intervention was needed), the IWT sector was marginalized. However, considering its inherent advantages, the need for systematic development of IWT sector was always felt which is evident from the fact that since independence, several Committees studied IWT system of the country from time to time and advocated systematic development of the mode.National Transport Policy Committee in its report (1980) accordingly recommended for setting up of an Authority for development and regulation of inland waterways,which led to formation of Inland Waterways Authority of India(IWAI)in 1986 for development and regulation of inland waterways.

IWAI undertakes infrastructure development and maintenance works on National Waterways.It also takes up techno-economic feasibility studies and prepares proposals for declaration of other waterways as National Waterways. It also advises Central Government on matters related to IWT and assists States in development of IWT sector.

As per constitutional provisions, only those waterways which are declared as National Waterways come under the purview of Central Government while rest of waterways remain in the purview of respective State Government.Since formation of IWAI, five waterways namely:

- Ganga
- Brahmaputra
- West Coast Canal with Udyogmandal and Champakara Canals
- Kakinada-Puducherry Canals system along with Godavari and Krishna rivers
- East Coast Canal with Brahmani river and Mahanadi delta have been declared as National Waterways.
- One more waterway namely Barak river is under consideration of the Central Government for declaration as a National Waterways.
- Waterways Identified by National Transport Policy Committee for consideration for declaration as National Waterways
- The National Transport Policy Committee (1980) recommended the following principles for declaration of a national waterway.
- It should possess capability of navigation by mechanically propelled vessels of a reasonable size.
- It should have about 45 m wide channel and minimum 1.5m depth.
- It should be a continuous stretch of 50 kms.The only exception to be made to waterway length is for urban conglomerations and intra-port traffic.
- It should pass through and serve the interest of more than one State (or).
- It should connect a vast and prosperous hinterland and Major Ports (or).

- It should pass through a strategic region where development of navigation is considered necessary to provide logistic support for national security (or).
- It should connect places not served by any other modes of transport.
- The National Transport Policy Committee considered the following waterways as having the potential for declaration as national waterway
- The Sunderbans
- The Mahanadi
- The Narmada
- The Mandovi,Zuari rivers and Cumberjua Canal in Goa
- The Tapi,

Hydrographic surveys and techno economic feasibility studies are the Pre-requisites for establishing the potential and viability of a waterway. Extensive surveys and investigations have been carried out on all the above waterways based on which three waterways have been so far declared as national waterways namely the Ganga, the Brahmaputra and the West Coast Canal. Development of many more new waterways as national waterways are planned during the 9th Plan period.

Background

Out of the 108 layers invisaged under project India-WRIS, 'Inland waterways' is one of the layer. India has about 14,500 km of navigable waterways, which comprise of rivers, canals, backwaters, creeks, etc. About 55 million tones of cargo are being moved annually by Inland Water Transport (IWT), in a fuel-efficient and environment-friendly mode. Its operations are currently restricted to a few stretches in the Ganga-Bhagirathi-Hooghly River, the Brahmaputra, the Barak River, the rivers in Goa, the backwaters in Kerala, inland waters in Mumbai and the deltaic regions of the Godavari-Krishna rivers. Besides these organized operations by mechanized vessels, country boats of various capacities also operate in various rivers and canals. Substantial quantum of cargo and passengers are transported in this unorganized sector as well. The Ganga-Bhagirathi-Hooghly between Allahabad-Haldia (1620 km) in UP, Bihar, Jharkhand and West Bengal, the Sadiya-Dhubri stretch of river Brahmaputra (891 km) in Assam and the Kollam-Kottapuram stretch of West Coast Canal along with Champakara and Udyogmandal Canals (205 km) in Kerala have so far been declared as national Waterways and are being developed for navigation by IWAI.

Indian Waterways Authority of India (IWAI) has completed hydrographic survey in five sectors known as NW-1, NW-2, NW-3, NW-4 and NW-5.Brief description of these stretches is given here.Figure:1 shows the index map of all the waterways.

- Development of National Waterways
- Three basic IWT related infrastructure for development of waterways are:
- Fairway or navigational channel with desired width and depth.
- Navigational aids for safe navigation and
- Terminals for berthing of vessels, loading/unloading of cargo and for providing interface with road and rail.

The fourth component for operationalising a viable IWT system is 'inland vessels' for transportation of goods and passengers. It is envisaged that once the fairway, terminals and navigational aids are provided to a threshold level, private sector investment in inland vessels, will increase dictated by market forces resulting in increase in inland fleet.Various projects for providing/maintaining fairway, terminals and navigational aids are being executed on National Waterways.

National Waterways No. 1 and 2 are typical alluvial rivers with characteristics of braiding, meandering and large water level fluctuation (both horizontal and vertical) between summer and monsoon months.On these rivers, several shallow areas (shoals) come up during low water season and maintenance of 2m Least Available Depth, particularly in upper reaches, becomes a difficult task. In these rivers,conservancy works (dredging and bandalling) are to be repeated every year on the shoals since after every monsoon the shoals are to be identified afresh and corrective measures (River Conservancy works) taken up. NW-3 on the other hand, is a tidal canal with predictable and uniform tidal variation in water level. On this waterway, therefore, once the desired depth is provided by capital dredging, it can be maintained for a number of years by undertaking nominal maintenance dredging from time to time as per actual requirement. NW-4 and 5 consist of both canal and river stretches. While canal portions need to be extensively dredged once to provide depth, on Godavari and Krishna rivers yearly dredging will be required and on Brahmani river five barrages with navigational locks have been proposed. Works of NW-4 and 5 will commence after completion of their DPR's, approval of development projects by competent authorities and allocation of adequate funds by the Government of India. PIB note for development of NW-4 has already been prepared and submitted to Ministry of Shipping for approval.

Some of the other waterways identified for development during the 9th Five Year plan and beyond are as follows:

Sunderbans

IWT route from Haldia to Bangladesh runs through the Sunderbans delta. This is a tidal inland waterway route which forms part of the trade and transit route under the Indo-Bangladesh Protocol. The hydrographic survey was conducted in 1988-89 and proposal submitted to the Government for its declaration as National Waterway.

Goa Waterways

Goa waterways are among the ten waterways identified by the National Transport Policy Committee (NTPC). These waterways comprises of Mondovi, Zuari rivers and Cumberjua canal having a length of about 122 kms. The waterways are fully functional with about 15 million tonnes annual cargo movement which mainly consists of iron ore from mines to Marma goa Port.

DVC Canal

This is a navigation canal system connecting Durgapur with Tribeni in Hooghly river which runs through a distance of 136 kms. Bhagwati Committee had identified this as a prospective waterway to be revived. Techno economic studies and the hydrographic surveys of this navigational canal system is in progress. Revival of this canal for movement of Coal from Durgapur mines to various destinations in NW-1 and 2 was proposed during the 9th Five Year Plan.

Narmada River

The Narmada river is the largest river in India flowing from East to West. It's source is near Amarkantak in the central river flows almost straight course towards to the Gulf of Khambot. River Narmada is one of the 10 major waterways identified for declaration as National Waterways by the National Transport Policy Committee(NTPC). The Pre-feasibility study for navigation on river Narmada waterway was carried out in 1981/ 82 by the Government of Gujarat. Development of Narmada in the lower reaches which are partly tidal below Sardar Sarovar Dam, and development of navigation in the reservoir area of the high dams is proposed during the 9th Five Year Plan.

Extension of National Waterway No-III

Based on the Techno Economic Feasibility study the stretch of the West Coast Canal from Kottapuram to Kollam was declared as National Waterway in 1993.

Extension of the waterway towards North up to Kasargod and towards South up to Kovalam is being persistently sought by public representatives and the State Government in view of the inadequacy of transport infrastructure in Kerala.

Other Waterways

In addition to above other waterways which are proposed (in the working group report) to be developed during the 9th Five Year Plan include Yamuna u/s Allahabad including the stretch in the Delhi region, IWT route at Mumbai to decongest the Mumbai city, and feeder routes in the existing National Waterways, waterways of Punjab and Haryana, Polavaram canal, Rajasthan canal, Mizoram waterways etc. Development of these waterways would be taken up either as central schemes or as centrally sponsored schemes where the State Govt.would have a major role to play.

National Waterways-1

Ganga-Bhagirathi-Hooghly river system from Allahabad to Haldia was declared as National Waterway No.1 vide National Waterway (Allahabad-Haldia stretch of the Ganga Bhagirathi- Hooghly river) Act 1982 (49 of 1982). It became operative from 27th Oct 1986 after the formation of the IWAI. The waterway extends from Haldia to Allahabad for a distance of 1620 kms.

The Hooghly river portion of the waterway from Haldia to Nabadwip is tidal. Sea going vessels navigate up to Calcutta (140 kms) and the fairway up to Calcutta is maintained by the Calcutta Port Trust. From Calcutta up to Tribeni there is no restrictions for navigation by inland vessels of a loaded draft up to 4m. From Nabadwip to Jangipur the waterway is formed by Bhagirathi river. Bhagirathi river is a regulated river because of the Barrages at Farakka and Jangipur.With the controlled discharge from the Farakka Barrage and limited river conservancy work a navigable depth of 2m is maintained in this route throughout the year.

Passenger Jetty-2 at Lot8

From Farakka upstream the navigable route is through the main Ganga river. The Feeder Canal and the navigation lock at Farakka become the link between the Bhagirathi and main Ganga upstream Farakka Barrage. The large variation in discharge along with unstable morphological condition of bank and bed, heavy sediment load,continuous braiding and meandering make development of navigational channel acomplex task.

NW-1 stretch is defined as Ganga-Bhagirathi-Hooghly river system from Allahabad to Haldia (1620km).This waterway is declared as National Waterway in 1986.NW-1 is passing through UP, Bihar, Jharkhand and West Bengal.Length of Haldia–Farakka waterway is 560Km, of Farakka–Patna is 460Km and of Patna–Allahabad is 600 km. Figure:1 shows the index map of National Waterway–1.

FAIRWAY DEVELOPMENT

The stretch between Haldia and Tribeni (196 km) is tidal and the least Available Depth (LAD) of more than 2 m is maintained naturally therein. During 2008-09, to maintain LAD in stretch upstream of Tribeni, 3000 m of bandals were erected and maintained in Tribeni-Farakka (364 km),and 15,705 m in Farakka-Allahabad (1060 km) stretches. In addition, 66,000 m3 of dredging was carried out in Tribeni-Farakka stretch and 1,02,000 m3 in Farakka–Varanasi stretch by deploying three Cutter Suction Dredgers (CSD) which are owned by IWAI. LAD of 2.5m was maintained between Haldia and Farakka (560 km) for about 330 days. LAD of 1.8 to 2m for about 300 days was maintained between Farakka and Patna (460 km). 1.5 m LAD between Patna and Varanasi (363 km) was maintained for about 330 days. However, between Varanasi-Allahabad, LAD of 1.2 m only could be maintained for about 300 days since proposed contract dredging could not be carried out because of unsatisfactory response to the dredging tender. Construction of two CSD units [a unit comprises one Cutter Suction Dredger (CSD), one Work Boat (WB) and one Accommodation Boat (AB)] progressed during the year through different shipbuilders who were awarded the work of construction and supply of these vessels.Both the ABs were supplied during 2008-09 while CSDs and WBs are expected to be delivered in 2009-10.

Terminals

For construction of high-level jetty at Patna, CPWD was conveyed the sanction of revised cost estimate of Rs 29.93 cr. CPWD awarded the work and construction work is in progress. They have been requested to complete the work by June 2010. Government of Bihar was requested for acquiring 1 acre of land for this high level jetty. At this site, the construction of low-level jetty was already completed during 2007-08. For construction of permanent terminal at GR jetty, IWAI had entrusted the work to CPWD on deposit basis. During 2008-09, CPWD awarded the work to the construction contractor and mobilization of machinery etc was done. For construction of permanent terminal at Varanasi, IWAI had entrusted the work to M/s. EPI Ltd on deposit basis and estimate/drawing/design etc were prepared by them. For a permanent terminal at Haldia also, alternative sites were inspected and it was decided in principle that its construction and management may be taken up on PPP mode through Project Development

Organisation/Project Development Committee (PDO/PDC) mechanism, IWAI has with M/s.IL&FS–IDC.

Besides these fixed terminals, floating terminals also exist at Haldia, Kolkata, Shantipur, Farakka, Rajmahal, Sahibganj, Manihari, Bhagalpur, Semaria, Patna, Buxar, Kaithy, Chunar and Allahabad. These terminals were maintained and used for transportation of cargo. For upgrading floating terminals, construction of twelve terminal pontoons for Haldia, Diamond Harbour, Kolkata, Shantipur, Katwa and Farakka, work has been completed.

Navigational Aids

Temporary channel marks for day navigation were erected and maintained between Tribeni and Allahabad all round the year. Besides, fortnightly thalweg surveys during low water period and monthly during flood period were carried out, river notices issued and pilotage provided to the cargo vessels. Night navigation aids (351 navigational lights fitted over country boats) were also maintained between Tribeni and Farakka (364 km). Tender for navigational lights fitted over country boats in Farakka-Patna stretch were also invited. To supplement these aids and for providing state of art 24 hrs navigation aids in the entire waterway, DGPS stations are proposed to be set up at Katwa, Bhagalpur, Patna and Varanasi. For this, civil work at Bhagalpur has been completed and DGPS instruments delivered during 2008-09. For trestle towers, supply of buoys and lights etc works were awarded and were under progress. Actions for taking over land for setting up of DGPS stations at Katwa, Patna and Varanasi were also initiated.

Facilities for safe 24x7 navigation

IWAI is also regularly issuing fortnightly river notices for the entire stretch of NW-1 (monthly during the flood season).24 hour Night navigation facilities are in place upto Farakka and this is to be extended upto varanasi by March 2010.A new DGPS station has been commissioned by December ,2010.Once they are commissioned,all navigational aids necessary for safe navigation would be in place on NW-1.These facilities would considerably reduce the turn around time for vessels.

11

RESERVOIR PLANNING

A reservoir or an artificial lake is used to store water. Reservoirs may be created in river valleys by the construction of a dam or may be built by excavation in the ground or by conventional construction techniques such a brickwork or cast concrete. The term reservoir may also be used to describe underground reservoirs such as an oil or water well.

A dam constructed in a valley relies on the natural topography to provide most of the basin of the reservoir. Dams are typically located at a narrow part of a valley downstream of a natural basin. The valley sides act as natural walls with the dam located at the narrowest practical point to provide strength and the lowest practical cost of construction. In many reservoir construction projects people have to be moved and re-housed, historical artifacts moved or rare environments relocated. Examples include the temples of Abu Simbel (which were moved before the construction of the Aswan Dam to create Lake Nasser from the Nile in Egypt) and the re-location of the village of Capel Celyn during the construction of Llyn Celyn.

Construction of a reservoir in a valley will usually necessitate the diversion of the river during part of the build often through a temporary tunnel or by-pass channel.

In hilly regions reservoirs are often constructed by enlarging existing lakes. Sometimes in such reservoirs the new top water level exceeds the watershed height on one or more of the feeder streams such as at Llyn Clywedog in Mid Wales. In such cases additional side dams are required to contain the reservoir.

Where the topography is poorly suited to a single large reservoir, a number of smaller reservoirs may be constructed in a chain such as in the River Taff valley where the three reservoirs Llwyn-on Reservoir, Cantref Reservoir and Beacons Reservoir form a chain up the valley.

BANK-SIDE RESERVOIR

Where water is taken from a river of variable quality or quantity, bank-side reservoirs may be constructed to store the water pumped or siphoned from the river. Such reservoirs are usually built partly by excavation and partly by the construction of a complete encircling bund or embankment which may exceed 6 km in circumference.

Both the floor of the reservoir and the bund must have an impermeable lining or core, often made of puddled clay. The water stored in such reservoirs may have a residence time of several months during which time normal biological processes are able to substantially reduce many contaminants and almost eliminate any turbidity.

The use of bank-side reservoirs also allows a water abstraction to be closed down for extended period at times when the river is unacceptably polluted or when flow conditions are very low due to drought. The London water supply system is one example of the use of bank-side storage for all the water taken from the River Thames and River Lee with many large reservoirs such as Queen Mary Reservoir visible along the approach to London Heathrow Airport.

Service reservoir

Service reservoirs store fully treated potable water close to the point of distribution. Many service reservoirs are constructed as water towers, often as elevated structures on concrete pillars where the landscape is relatively flat. Other service reservoirs are entirely underground, especially in more hilly or mountainous country. In the United Kingdom, Thames Water has many underground reservoirs in London built in the 1800s by the Victorians, most of which are lined with brick. Honor Oak Reservoir, which was completed in 1909, is believed to one of the largest of this type in Europe. The roof is supported on large brick pillars and arches and the outside surface is grassed over.

Service reservoirs perform several functions including ensuring sufficient head of water in the water distribution system and providing hydraulic capacitance in the system to even out peak demand from consumers enabling the treatment plant to run at optimum efficiency. Large service reservoirs can also be managed to so that energy costs in pumping are reduced by concentrating refilling activity at times of day when power costs are low. Five thousand years ago, the craters of extinct volcanoes in Arabia were used as reservoirs by farmers for their irrigation water. Dry climate and water scarcity in India led to early development of water management techniques, including the building of a reservoir at Girnar in 3000 BC. Artificial lakes dating to the 5th century BC have been found in ancient Greece. An artificial lake in present-day Madhya Pradesh province of India, constructed in the 11th century, covered 650 square metres (7,000 sq ft).

In Sri Lanka large reservoirs have been created by ancient Sinhalese kings in order to save the water for irrigation. The famous Sri Lankan king Par?kramab?hu I of Sri Lanka stated " do not let a drop of water seep into the ocean without benefiting mankind ". He created the reservoir named Parakrama Samudra(sea of King Parakrama), which has astonished archaeologists.

USES

Direct water supply

Many dammed river reservoirs and most bank-side reservoirs are used to provide the raw water feed to a water treatment plant which delivers drinking water through water mains. The reservoir does not simply hold water until it is needed; it can also be the first part of the water treatment process. The time the water is held for before it is released is known as the retention time. This is a design feature that allows particles and silts to settle out, as well as time for natural biological treatment using algae, bacteria and zooplankton that naturally live within the water.

However natural limnological processes in temperate climate lakes produces temperature stratification in the water body which tends to partition some elements such as manganese and phosphorus into deep, cold anoxic water during the summer months. In the autumn and winter the lake becomes fully mixed again. During drought conditions, it is sometimes necessary to draw down the cold bottom water and the elevated levels of manganese in particular can cause problems in water treatment plants.

Hydroelectricity

A reservoir generating hydroelectric includes turbines connected to the retained water body by large diameter pipes. These generating sets may be at the base of the dam or some distance away. Some reservoirs generating hydro-electricity use pumped re-charge in which a high level reservoir is filled with water using high performance electric pumps at times when electricity demand is low and then uses this stored water to generate electricity by releasing the stored water into a low level reservoir when electricity demand is high. Such systems are called pump storage schemes.

Controlling watercourses

Reservoirs can be used in a number of ways to control how water flows through downstream waterways.

Downstream water supply - water may be released from an upland reservoir so that it can be abstracted for drinking water lower down the system, sometimes hundred of miles further down downstream

Irrigation - water in an irrigation reservoir may be released into networks of canals for use in farmlands or secondary water systems. Irrigation may also be supported by reservoirs which maintain river flows allowing water to be abstracted for irrigation lower down the river.

Flood control - also known as an "attenuation" or "balancing" reservoir, flood control reservoirs collect water at times of very high rainfall, then release it slowly over the course of the following weeks or months. Some of these reservoirs are constructed across the river line with the onward flow controlled by an orifice plate. When river flow exceeds the capacity of the orifice plate water builds behind the dam but as soon as the flow rate reduces the water behind the dam slowly releases until the reservoir is empty again. In some cases such reservoirs only function a few times in a decade and the land behind the reservoir may be developed as community or recreational land. A new generation of balancing dams are being developed to combat the climatic consequences of climate change. They are called "Flood Detention Reservoirs". Because these reservoirs will remain dry for long periods, there may be a risk of the clay core drying out reducing its structural stability. Recent developments include the use of composite core fill made from recycled materials as an alternative to clay.

Canals - Where a natural watercourse's water is not available to be diverted into a canal, a reservoir may be built to guarantee the water level in the canal; for example, where a canal climbs to cross a range of hills through locks.

Recreation - water may be released from a reservoir to artificially create or supplement white-water conditions for kayaking and other white-water sports. On salmonid rivers special releases (in Britain called freshets) are made to encourage natural migration behaviours in fish and to provide a variety of fishing conditions for anglers.

FLOW BALANCING

Occasionally reservoirs are used to balance the flow in highly managed systems, taking in water during high flows and releasing it again during low flows. In order for this to work without pumping requires careful control of water levels using adjustable sluices. One example of this is Llyn Tegid in North Wales. This is a natural lake whose level was raised by a low dam and into which the River Dee flows or discharges depending upon flow conditions at the time as part of the River Dee regulation system. This mode of operation is a form of hydraulic capacitance in the river system.

Recreation

The water bodies provided by many reservoirs often allow some recreational uses such as fishing, boating, and other activities. Special rules may apply for the safety of the public and to protect the quality of the water and the ecology of the surrounding area. Many reservoirs now support and encourage less informal and less structured recreation such as natural history, bird watching, landscape painting, walking and hiking and often provide information boards and interpretation material to encourage responsible use.

Operation

Water falling as rain upstream of the reservoir together with any groundwater emerging as springs is stored in the reservoir. Any excess water can be spilled via a specifically designed spillway. Stored water may be piped by gravity for use as drinking water, to generate hydro-electricity or to maintain river flows to support downstream uses. Occasionally reservoirs can be managed to retain high rain-fall events to prevent or reduce downstream flooding. Some reservoirs support several uses and the operating rules may be complex. Most modern reservoirs have a specially designed draw-off tower that can discharge water from the reservoir at different levels both to access water as the reservoir draws down but also to allow water of a specific quality to be discharged into the downstream river as compensation water.

The operators of many upland or in-river reservoirs have obligations to release water into the downstream river to maintain river quality, support fisheries, maintain downstream industrial uses. maintain recreational use or for a range of other requirements. Such releases are known as compensation water.

Terminology

The terminology for reservoirs varies from country to country. In most of the world reservoir areas are expressed in km2 whilst in the USA acres are commonly used. For volume either m3 or km3 are widely used with acre feet used in the USA.

The capacity, volume or storage of a reservoir is usually divided into distinguishable areas. Dead or inactive storage refers to water in a reservoir that cannot be drained by gravity through a dam's outlet works, spillway or power plant intake and can only be pumped out. Dead storage allows sediments to settle which improves water quality and also creates hydraulic head along with an area for fish during low levels. Active or live storage is the portion of the reservoir that can be utilized for flood control, power production, navigation and downstream releases.

In addition, a reservoir's flood control capacity is the amount of water it can regulate during flooding. The surcharge capacity is the capacity of the reservoir above the spillway crest that cannot be regulated. In the United States the water below the normal maximum level of a reservoir is called the conservation pool. In the UK top water level describes the reservoir full state whist fully drawn down describes the minimum retained volume.

MODELLING RESERVOIR MANAGEMENT

There is a wide variety of software for modelling reservoirs, from the specialist Dam Safety Program Management Tools (DSPMT) to the relatively simple WAFLEX, to integrated models like the Water Evaluation And Planning system (WEAP) that place reservoir operations in the context of system-wide demands and supplies.

Safety

In many countries large reservoirs are closely regulated to try and prevent or minimise failures of containment. Whilst much of the effort is directed at the dam and its associated structures as the weakest part of the overall structure, the aim of such controls is to prevent an uncontrolled release of water from the reservoir. Reservoir failures can generate huge increases in flow down a river valley with the potential to wash away towns and villages and cause considerable loss of life such as the devastation following the failure of containment at Llyn Eigiau which killed 17 people.

A notable case of reservoirs being used as an instrument of War involved the British Royal Air Force Dambusters raid on Germany in World War II (codenamed "Operation Chastise"), in which three German reservoir dams were selected to be breached in order to impact on German infrastructure and manufacturing and power capabilities deriving from the Ruhr and Eder rivers. The economic and social impact was derived from the enormous volumes of previously stored water that swept down the valleys wreaking destruction. This raid later became the basis for several films.

All reservoirs will have a monetary cost/benefit assessment made before construction to see if the project is worth proceeding with. However, such analysis can often omit the environmental impacts of dams and the reservoirs that they contain. Some impacts such as the greenhouse gas production associated with concrete manufacture are relatively easy to estimate. Other impact on the natural environment and social and cultural effects can be more difficult to assess and to weigh in the balance but identification and quantification of these issues are now commonly required in major construction projects in the developed world.

Climate Change

Depending upon the circumstances, a reservoir built for hydro-electricity generation can either reduce or increase the net production of greenhouse gases. An increase can occur if plant material in the flooded areas decays in an anaerobic environment releasing (methane and carbon dioxide). This apparently counter intuitive position arises because much carbon is released as methane which is approximately 8 times more potent as a greenhouse gas than carbon dioxide.

A study for the National Institute for Research in the Amazon found that Hydroelectric reservoirs release a large pulse of carbon dioxide from above-water decay of trees left standing in the reservoirs, especially during the first decade after closing. This elevates the global warming impact of the dams to levels much higher than would occur by generating the same power from fossil fuels.

According to the World Commission on Dams report (Dams And Development), when the reservoir is relatively large and no prior clearing of forest in the flooded area was undertaken, greenhouse gas emissions from the reservoir could be higher than those of a conventional oil-fired thermal generation plant. For instance, In 1990, the impoundment behind the Balbina Dam in Brazil (inaugurated in 1987) had over 20 times the impact on global warming than would generating the same power from fossil fuels, due to the large area flooded per unit of electricity generated.

A decrease can occur if the dam is used in place of traditional power generation, since electricity produced from hydroelectric generation does not give rise to any flue gas emissions from fossil fuel combustion (including sulfur dioxide, nitric oxide and carbon monoxide from coal). The Tucurui dam in Brazil (closed in 1984) had only 0.4 times the impact on global warming than would generating the same power from fossil fuels. Dams can produce a block for migrating fish, trapping them in one area, producing food and a habitat for various water-birds. They can also flood various ecosystems on land and may cause extinctions.

Human Impact

Dams can severely reduce the amount of water from reaching countries downstream of them, causing water stress between the countries, e.g. the Sudan and Egypt, which damages farming businesses in the downstream countries, and reduces drinking water.

Farms and villages, e.g. Ashopton can be flooded by the creation of reservoirs, ruining many livelihoods. For this very reason, worldwide 80 million people have had to be forcibly relocated due to dam construction.

Limnology

The limnology of reservoirs has many similarities to that of lakes of equivalent size. There are however significant differences. Many reservoirs experience considerable variations in level producing significant areas that are intermittently underwater or dried out. This greatly limits the productivity or the water margins and limits the number of species able to survive in these conditions.

Upland reservoirs tend to have a much shorter residence time than natural lakes and this can lead to more rapid cycling of nutrients through the water body so that they are more quickly lost to the system. This may be seen as a mismatch between water chemistry and water biology with a tendency for the biological component to be more oligotrophic than the chemistry would suggest.

Conversely, lowland reservoirs drawing water from nutrient rich rivers, may show exaggerated eutrophic characteristics because the residence time in the reservoir is much greater than in the river and the biological systems have a much greater opportunity to utilise the available nutrients.

Deep reservoirs with multiple level draw off towers can discharge deep cold water into the downstream river greatly reducing the size of any hypolimnion. This in turn can reduce the concentrations of phosphorus released during any annual mixing event and may therefore reduce productivity.

The Dams in front of reservoirs act as knickpoints-the energy of the water falling from them reduces and deposition is a result below the Dams.

Seismicity

The filling (impounding) of reservoirs has often been attributed to reservoir-triggered seismicity (RTS) as seismic events have occurred near large dams or within their reservoirs in the past. These events may have been triggered by the filling or operation of the reservoir and are on a small scale when compared to the amount of reservoirs worldwide. Of over 100 recorded events, early examples include the 60 m (197 ft) tall Marathon Dam in Greece (1929), the 221 m (725 ft) tall Hoover Dam in the U.S. (1935).

Most events involve large dams and small amounts of seismicity. The only four recorded events above a 6.0- magnitude (Mw) are the 103 m (338 ft) tall Koyna Dam in India which registered a Mw of 6.3 along with the 120 m (394 ft) Kremasta Dam in Greece which registered a 6.3-Mw as well. Following those two, the next largest were the 122 m (400 ft) high Kariba Dam in Zambia at 6.25-Mw and the 105 m (344 ft) Xinfengjiang Dam in China at 6.1-Mw.

Disputes occur over when RTS has occurred due to a lack of hydrogeological knowledge at the time of the event. It is accepted though that the infiltration of water into pores and the weight of the reservoir do contribute to RTS patterns. For RTS to occur, there must be a seismic structure near the dam or its reservoir and the seismic structure must be close to failure. Additionally, water must be able to infiltrate the deep rock stratum as the weight of a 100 m (328 ft) deep reservoir will have little impact when compared the deadweight of rock on a crustal stress field which may be located at a depth of 10 km (6 mi) or more.

Micro climate

Reservoirs may change the local micro-climate increasing humidity and reducing extremes of temperature. Such effects are claimed by some South Australian winerys as increasing the quality of the wine production.

Sedimentation

Sedimentation is the tendency for particles in suspension to settle out of the fluid in which they are entrained, and come to rest against a barrier. This is due to their motion through the fluid in response to the forces acting on them: these forces can be due to gravity, centrifugal acceleration or electromagnetism.

In geology sedimentation is often used as the polar opposite of erosion, i.e., the terminal end of sediment transport. In that sense it includes the termination of transport by saltation or true bedload transport. Settling is the falling of suspended particles through the liquid, whereas sedimentation is the termination of the settling process.

Sedimentation may pertain to objects of various sizes, ranging from large rocks in flowing water to suspensions of dust and pollen particles to cellular suspensions to solutions of single molecules such as proteins and peptides. Even small molecules such as aspirin can be sedimented, although it can be difficult to apply a sufficiently strong force to produce significant sedimentation.

The term is typically used in geology, to describe the deposition of sediment which results in the formation of sedimentary rock, and in various chemical and environmental fields to describe the motions of often-smaller particles and molecules. Process is also used in biotech industry to separate out cells from the culture media.

In a sedimentation experiment called tripothsis, the applied force accelerates the particles to a terminal velocity vterm at which the applied force is exactly canceled by an opposing drag force. For small enough particles (low Reynolds number), the drag force varies linearly with the terminal velocity, i.e., Fdrag = fvterm (Stokes flow) where f depends only on the properties of the particle and the surrounding fluid. Similarly, the applied force generally varies linearly with some coupling constant (denoted here as q) that depends only on the properties of the particle, Fapp = qEapp. Hence, it is generally possible to define a sedimentation coefficient that depends only on the properties of the particle and the surrounding fluid. Thus, measuring s can reveal underlying properties of the particle.

In many cases, the motion of the particles is blocked by a hard boundary; the resulting accumulation of particles at the boundary is called a sediment. The concentration of particles at the boundary is opposed by the diffusion of the particles.

The sedimentation of particles under gravity is described by the Mason-Weaver equation, which has a simple exact solution. The sedimentation coefficient s in this case equals mb / f, where mb is the buoyant mass.

The sedimentation of particles under the centrifugal force is described by the Lamm equation, which likewise has an exact solution. The sedimentation coefficient s also equals mb / f, where mb is the buoyant mass. However, the Lamm equation differs from the Mason-Weaver equation because the centrifugal force depends on radius from the origin of rotation, whereas gravity is presumed constant. The Lamm equation also has extra terms, since it pertains to sector-shaped cells, whereas the Mason-Weaver equation pertains to box-shaped cells (i.e., cells whose walls are aligned with the three Cartesian axes).

Particles with a charge or dipole moment can be sedimented by an electric field or electric field gradient, respectively. These processes are called electrophoresis and dielectrophoresis, respectively. For electrophoresis, the sedimentation coefficient corresponds to the particle charge divided by its drag (the electrophoretic mobility). Similarly, for dielectrophoresis, the sedimentation coefficient equals the particle's electric dipole moment divided by its drag.

In geology, sedimentation is the deposition of particles carried by a fluid flow. For suspended load, this can be expressed mathematically by the Exner equation, and results in the formation of depositional landforms and the rocks that constitute sedimentary record. An undesired increased transport and sedimentation of suspended material is called siltation, and it is a major source of pollution in waterways in some parts of the world. Climate change also affect siltation rates.

Water quality

Water quality is the physical, chemical and biological characteristics of water. It is a measure of the condition of water relative to the requirements of one or more biotic species and or to any human need or purpose. It is most frequently used by reference to a set of standards against which compliance can be assessed. The most common standards used to assess water quality relate to drinking water, safety of human contact and for the health of ecosystems.

In the setting of standards, agencies make political and technical/scientific decisions about how the water will be used. In the case of natural water bodies, they also make some reasonable estimate of pristine conditions. Different uses raise different concerns and therefore different standards are considered. Natural water bodies will vary in response to environmental conditions. Environmental scientists work to understand how these systems function which in turn helps to identify the sources and fates of contaminants. Environmental lawyers and policy makers work to define legislation that ensure that water is maintained at an appropriate quality for its identified use.

The vast majority of surface water on the planet is neither potable nor toxic. This remains true even if sea water in the oceans (which is too salty to drink) is not counted. Another general perception of water quality is that of a simple property that tells whether water is polluted or not. In fact, water quality is a very complex subject, in part because water is a complex medium intrinsically tied to the ecology of the Earth. Industrial pollution is a major cause of water pollution, as well as runoff from agricultural areas, urban stormwater runoff and discharge of treated and untreated sewage (especially in developing countries).

Categories

The parameters for water quality are determined by the intended use. Work in the area of water quality tends to be focused on water that is treated for human consumption or in the environment.

Human consumption

Contaminants that may be in untreated water include microorganisms such as viruses and bacteria; inorganic contaminants such as salts and metals; organic chemical contaminants from industrial processes and petroleum use; pesticides and herbicides; and radioactive contaminants. Water quality depends on the local geology and ecosystem, as well as human uses such as sewage dispersion, industrial pollution, use of water bodies as a heat sink, and overuse (which may lower the level of the water).

In the United States, the U.S. Environmental Protection Agency (EPA) limits the amounts of certain contaminants in tap water provided by public water systems. The Safe Drinking Water Act authorizes EPA to issue two types of standards: primary standards regulate substances that potentially affect human health, and secondary standards prescribe aesthetic qualities, those that affect taste, odor, or appearance.

The U.S. Food and Drug Administration (FDA) regulations establish limits for contaminants in bottled water that must provide the same protection for public health. Drinking water, including bottled water, may reasonably be expected to contain at least small amounts of some contaminants. The presence of these contaminants does not necessarily indicate that the water poses a health risk.

Some people use water purification technology to remove contaminants from the municipal water supply they get in their homes, or from local pumps or bodies of water. For people who get water from a local stream, lake, or aquifer (well), their drinking water is not filtered by the local government.

Environmental water quality

Environmental water quality, also called ambient water quality, relates to water bodies such as lakes, rivers, and oceans. Water quality standards vary significantly due to different environmental conditions, ecosystems, and intended human uses.

Toxic substances and high populations of certain microorganisms can present a health hazard for non-drinking purposes such as irrigation, swimming, fishing, rafting, boating, and industrial uses. These conditions may also affect wildlife which use the water for drinking or as a habitat. Modern water quality laws general specify protection of fisheries and recreational use and require as a minimum,retention of current quality standards.

There is some desire among the public to return water bodies to pristine, or pre-industrial conditions. Most current environmental laws focus of the designation of uses. In some countries these allow for some water contamination as long as the particular type of contamination is not harmful to the designated uses. Given the landscape changes in the watersheds of many freshwater bodies, returning to pristine conditions would be a significant challenge. In these cases, environmental scientists focus on achieving goals for maintaining healthy ecosystems and may concentrate on the protection of populations of endangered species and protecting human health.

Measurement

The complexity of water quality as a subject is reflected in the many types of measurements of water quality indicators. Some of the simple measurements listed below can be made on-site - temperature, pH, dissolved oxygen, conductivity, Oxygen Reduction potential (ORP), turbidity, Secchi Disk depth- in direct contact with the water source in question. More complex measurements that must be made in a lab setting require a water sample to be collected, preserved, and analyzed at another location.

Making these complex measurements can be expensive. Because direct measurements of water quality can be expensive, ongoing monitoring programs are typically conducted by government agencies. However, there are local volunteer programs and resources available for some general assessment. Tools available to the general public are on-site test kits commonly used for home fish tanks and biological assessments.

Testing in response to natural disasters and other emergencies

Inevitably after events such as earthquakes and Tsunamis, there is an immediate response by the aid agencies as relief operations get underway to try and restore basic infrastructure and provide the basic fundamental items that are necessary for survival and subsequent recovery. Access to clean drinking water and adequate sanitation is a priority at times like this. The threat of disease increases hugely due to the large numbers of people living close together, often in squalid conditions, and without proper sanitation.

After a natural disaster, as far as water quality testing is concerned there are widespread views on the best course of action to take and a variety of methods can be employed. The key basic water quality parameters that need to be addressed in an emergency are bacteriological indicators of fecal contamination, Free Chlorine Residual, pH, turbidity and possibly Conductivity/TDS. There are a number of portable water test kits on the market widely used by aid and relief agencies for carrying out such testing.

The following is a list of indicators often measured by situational category:

Drinking water

Filtering water

- Alkalinity
- Color of water
- pH
- Taste and odor (geosmin, 2-methylisoborneol (MIB), etc)
- Dissolved metals and salts (sodium, chloride, potassium, calcium, manganese, magnesium)
- Microorganisms such as fecal coliform bacteria (Escherichia coli), Cryptosporidium, and Giardia lamblia
- Dissolved metals and metalloids (lead, mercury, arsenic, etc.)
- Dissolved organics: colored dissolved organic matter (CDOM), dissolved organic carbon (DOC)
- Radon
- Heavy metals
- Pharmaceuticals
- Hormone analogs

Environmental

Chemical assessment

- (also see salinity)
- Dissolved Oxygen (DO)
- nitrate-N
- orthophosphates
- Chemical oxygen demand (COD)
- Biochemical oxygen demand (BOD)
- Pesticides

Physical assessment

- pH
- Temperature

- Total suspended solids (TSS)
- Turbidity

BIOLOGICAL ASSESSMENT

Biological monitoring metrics have been developed in many places, and one widely used measure is the presence and abundance of members of the insect orders Ephemeroptera, Plecoptera and Trichoptera. (Common names are, respectively, Mayfly, Stonefly and Caddisfly.) EPT indexes will naturally vary from region to region, but generally, within a region, the greater the number of taxa from these orders, the better the water quality. EPA and other organizations in the United States offer guidance on developing a monitoring program and identifying members of these and other aquatic insect orders.

Individuals interested in monitoring water quality who cannot afford or manage lab scale analysis can also use biological indicators to get a general reading of water quality. One example is the IOWATER volunteer water monitoring program, which includes a benthic macroinvertebrate indicator key.

STANDARDS AND REPORTS

Canada

In Canada, Manitoulin Streams Improvement Association has become a leading model for water quality and fisheries rehabilitation. The association partners with landowners, farmers, fishermen and the general public to improve water quality and the fisheries resource on Manitoulin Island and the Great Lakes. They do this by:

- Restricting livestock access to certain points on the river or installing alternative watering sources like nose pumps.
- Repair the Riparian Zone by planting trees and grasses to stabalize shorelines,provide habitat.
- Create in stream habitat to increase fish and invertebrate populations.

Since 2000, Manitoulin Streams has rehabilitated 23 major sites on 4 waterways. They have had a Class Environmental Assessment conducted on 184 waterways on Manitoulin Island. The report identified 10 priority waterways that needed to be rehabilitated. Manitoulin Streams has conducted work on 4 of the ten and has plans to work on a 5th, the Mindemoya River in the Summer of 2010.

Flood Storage

The components of online storage works normally include:

(a) an impounding structure - generally comprises an earth or concrete structure across the river and floodplain, behind which the water is stored;

(b) a flow control structure - normally located within the impounding structure to control the outflow from the storage area;

(c) a spillway - to pass floods that are more extreme than those that the reservoir is designed to attenuate.

The flow control structure can be a fixed throttle (such as an orifice, pipe or flume) sized to have little effect on normal flows, but requiring a significant rise in upstream water level to pass larger flows. Sometimes, the control structure incorporates gates, which are normally left open, but are operated during floods to ensure that downstream flows do not overtop the downstream flood defences.

Flood Storage Area

As noted above, an online flood storage reservoir must also be designed to deal safely with more extreme floods than it is designed to attenuate; these could otherwise lead to the water level upstream of the impounding structure rising above safe levels or unacceptable damage to the impounding structure or surrounding land. This normally involves the provision of an overflow weir or spillway to convey the extreme floods. Sometimes the entire impounding embankment - or a major part of it - provides the overflow route for floods, having been designed to withstand the flow velocities involved. In other cases, the flow control structure incorporates the necessary arrangements for passing flood flows, either through gates or over fixed weirs and other structures.

Offline flood storage works generally consist of:

(a) an intake structure to divert water to the storage area when the river flow or level exceeds a pre-determined value;

(b) a storage area that comprises a reservoir separated from the river, formed either by low ground levels (natural or excavated) or by retaining structures (embankments, walls, or a combination of the two);

(c) an outlet structure that returns water from the storage area to the river after the flood peak has passed;

(d) a spillway to pass floods that are more extreme than those that the reservoir is designed to attenuate.

The intake structure (which is of course an 'offtake' from the perspective of the river) often includes two components that may be located separately:

- a device designed to control flows passed downriver;
- an inflow structure to the storage area.

The device controlling downstream flow can take most of the forms used for the control of online flood storage works. Rather than building a new control structure, a 'natural' control may be used on the river in some cases, or a pre-existing control such as an existing weir in others. In these cases, the offline storage works are designed to make use of the existing stage/discharge characteristics of the river.

Weirs (Rickard et al, 2003) are often favoured for the inflow structure, sometimes in the form of a sideweir (May et al, 2003) along the riverbank. They have the advantage of beginning to operate whenever river levels rise above a given value, but need to be long to limit the rise in river level needed to pass greater flows into the storage area. For this reason, a gated arrangement is often used.

The offline flood storage area generally lies within the floodplain, but isolated by purpose-built walls or embankments (see Chapter 9). The volume available for storage in the reservoir depends on the water depth that can be obtained, which is controlled by existing ground levels, and the peak flood level that can be accommodated. This depth is often limited, making it necessary for the reservoir to cover a large area. Choosing a site where the ground is low - either naturally or as a result of excavations (for example, for gravel pits) - increases the depth available, but may mean that pumps are needed to empty the reservoir after the flood has passed.

The outlet may be by gravity (generally using gates), pumped or by a combination of the two (with gravity flow initially, then pumping to drain the lowest sections). The outlet capacity required depends on the volume stored and the time allowed for the system to be fully drained. Whenever the reservoir contains water - and so has less capacity to accommodate another flood - the standard of protection provided will be temporarily compromised. If the outfall is to a tidal river, it may only be possible to complete the emptying the reservoir at low tide, perhaps via a flapgate.

As for flood storage works, arrangements for the safe passage of extreme floods are again needed to protect the reservoir and its associated structures. These may form a major part of the facility, particularly if the inlet to the storage area is uncontrolled.

Seepage: steady state flow of water

If fluid pressures in a soil deposit are uniformly increasing with depth according to u = wgzw then hydrostatic conditions will prevail and the fluids will not be flowing through the soil. zw is the depth below the water table. However, if the water table is sloping or there is a perched water table as indicated in the accompanying sketch, then seepage will occur. For steady state seepage, the seepage velocities are not varying with time. If the water tables are changing levels with time, or if the soil is in the process of consolidation, then steady state conditions do not apply.

Darcy's law

Darcy's law states that the volume of flow of the pore fluid through a porous medium per unit time is proportional to the rate of change of excess fluid pressure with distance. The constant of proportionality includes the viscosity of the fluid and the intrinsic permeability of the soil. For the simple case of a horizontal tube filled with soil

The total discharge, Q (units of volume per time, e.g., ft^3/s or m^3/s) is proportional to the intrinsic permeability, K, the cross sectional area, A, and rate of pore pressure change with distance, , and inversely proportional to the dynamic viscosity of the fluid, ?. The negative sign is needed because fluids flow from high pressure to low pressure. So if the change in pressure is negative (in the x-direction) then the flow will be positive (in the x-direction). The above equation works well for a horizontal tube, but if the tube was inclined so that point b was a different elevation than point a, the equation would not work. The effect of elevation is accounted for by replacing the pore pressure by excess pore pressure, ue defined as:

ue = u wgz

where z is the depth measured from an arbitrary elevation reference (datum). Replacing u by ue we obtain a more general equation for flow:

Dividing both sides of the equation by A, and expressing the rate of change of excess pore pressure as a derivative, we obtain a more general equation for the apparent velocity in the x-direction: where vx has units of velocity and is called the Darcy velocity, or discharge velocity. The seepage velocity vsx = (average velocity of fluid molecules in the pores) is related to the Darcy velocity, and the porosity, n

Civil Engineers predominantly work on problems that involve water and predominantly work on problems on earth (in earths gravity). For this class of problems, civil engineers will often write Darcy's law in a much simpler form :

$v = ki$

where k is called permeability, and is defined as , and i is called the hydraulic gradient. The hydraulic gradient is the rate of change of total head with distance. The total head, h at a point is defined as the height (measured relative to the datum) to which water would raise in a piezometer at that point. The total head is related to the excess water pressure by:

$ue = wgh + \text{Constant}$

and the Constant is zero if the datum for head measurement is chosen at the same elevation as the origin for the depth, z used to calculate ue.

Typical values of permeability

Values of the permeability, k, can vary by many orders of magnitude depending on the soil type. Clays may have permeability as small as about , gravels may have permeability up to about . Layering and heterogeneity and disturbance during the sampling and testing process make the accurate measurement of soil permeability a very difficult problem.

Flow nets

Darcy's Law applies in one, two or three dimensions. In two or three dimensions, steady state seepage is described by Laplace's equation. Computer programs are available to solve this equation. But traditionally two-dimensional seepage problems were solved using and a graphical procedure known called flow nets.

One set of lines in the flow net are in the direction of the water flow (flow lines), and the other set of lines are in the direction of constant total head (equipotential lines). Flow nets may be used for example to estimate the quantity of seepage under dams and sheet piling. A plan flow net to estimate flow of water from a stream to a discharging well

Seepage forces and erosion

When the seepage velocity is great enough, erosion can occur because of the frictional drag exerted on the soil particles. Vertically upwards seepage is a source of danger on the downstream side of sheet piling and beneath the toe of a dam or levee. Erosion of the soil, known as "piping", can lead to failure of the structure and to sinkhole formation. Seeping water removes soil, starting from the exit point of the seepage, and erosion advances upgradient. The term sand boil is used to describe the appearance of the discharging end of an active soil pipe.

Seepage pressures

Seepage in an upward direction reduces the effective stress within the soil. When the water pressure at a point in the soil is equal to the total vertical stress at that point, the effective stress is zero and the soil has no frictional resistance to deformation. For a surface layer, the vertical effective stress becomes zero within the layer when the upward hydraulic gradient is equal to the critical gradient. At zero effective stress soil has very little strength and layers of relatively impermeable soil may heave up due to the underlying water pressures.

The loss in strength due to upward seepage is a common contributor to levee failures. The condition of zero effective stress associated with upward seepage is also called liquefaction, Quicksand, or a boiling condition. Quicksand was so named because the soil particles move around and appear to be 'alive' (the biblical meaning of 'quick' - as opposed to 'dead').

Reservoirs store surface water for use at a later time. Surface water reservoirs are created by building a dam or river at the outlet end of a LAKE or reservoir to store WATER. Reservoir water may be used for both off-stream (consumptive) uses such as irrigation, municipal and industrial; and for in-stream (nonconsumptive) uses such as power, recreational developments, FLOOD control and wildlife habitat improvement. Reservoirs may be used to improve the holding capacity of an existing lake or to create new lakes on rivers, creeks or coulees.

STORAGE CAPACITY

Reservoirs, as discussed here, do not include any type of subsurface reservoir structure that stores water, natural gas or oil. Surface water reservoirs are usually classified by 3 basic physical characteristics: maximum depth, flooded area and the volume of water stored at their full supply level (FSL). Large reservoirs may have several target storage levels, each level planned to ensure a usable amount of water at a certain time of year. For instance, in addition to its normal or FS level, a reservoir may have a conservation storage level (ie, the desirable level for a flood-control reservoir at the start of each flooding period), a flood storage level (the maximum desirable level that is permitted for reservoir safety during floods), a dead storage level (the level below which water may not be withdrawn for consumptive uses) and various other target levels established to meet such needs as summer recreational use, irrigation uses and hydropower discharges throughout the year.

Before construction, reservoir sites are subjected to detailed studies to determine their ability to meet the needs of potential users and to assess the overall benefits that can be derived from the project. Studies are usually carried out on a monthly basis, using extensive periods of estimated or recorded monthly historical streamflow to evaluate the project's performance under DROUGHT and flood conditions that have occurred historically. The results are used to determine if proposed reservoir would adequately serve its potential users if similar droughts occurred in the future.

A reservoir used for municipal purposes will be designed to have no shortages because the users (people in cities) cannot tolerate periods without water. When uses are such that shortages, although they may be inconvenient, will not cause severe economic hardship, reservoirs are often designed to allow for an acceptable percentage of shortages in critically dry years.

This permits more effective use to be made of the available water resources. Two or more reservoirs are often used to maximize the supply potential of a DRAINAGE BASIN. The potential improvement that might be achieved by adding additional potential reservoir storage sites to an existing system may be evaluated by hydrologic studies involving both the existing and the proposed reservoirs. The studies assume that they are being utilized in combination to optimize their overall effectiveness.

Evaporation

Water evaporates from the surface of reservoirs when they are not covered by ICE. These losses, known as net evaporation, may be estimated by subtracting measured or estimated precipitation from estimated evaporation on a reservoir's surface. The volume of reservoir losses caused by net evaporation may be calculated by multiplying net evaporation by a reservoir's surface area.

Water Cycle

The water cycle is a key element of change in the environment and is essential to all life (artwork by Michael Lee).

When hydrologic studies indicate that the addition of further storage capacity will not significantly increase the yield of a reservoir, either because the additional water stored would evaporate before being used or because all available water has been used, the reservoir has reached its optimum storage capacity and further storage increases would be wasted. It may by feasible to add storage for flood-control purposes. This alternative would require additional studies of potential flooding problems in the downstream river basin to determine if such storage is warranted.

Water Discharge

Reservoirs have spillways to divert excess water during above-normal periods. The spillway's capacity is based both on the economic consequences and on the potential danger to human life that might occur if the dam fails. When the potential exists for downstream loss of life, extensive property damage or both, spillways are usually designed to carry the largest flood that may be expected to occur at that site. After construction, reservoirs are operated to co-ordinate consumptive, recreational and power uses with downstream releases to maintain predetermined levels of minimum flow, thus establishing monthly reservoir levels, regulated discharge and diversions.

Environmental Effects

Several changes occur to the environment adjacent to a reservoir. The lake created by the reservoir provides an additional recreational area, but the pre-existing river is destroyed and the area at the river's edge that used to provide shelter to animals and birds is lost. The river downstream from the reservoir usually has lower peaks and higher low flows because of the moderating effect of the reservoirs during floods and because of additional water released during dry periods.

Colder water is released from the reservoir and fish must either adjust to the new environment or emigrate. However, there is often an increase in volume and diversity of the fish stocked in the reservoir itself, resulting in a net gain to the fishery.

The overall water quality in the reservoir and downstream tends to improve because the reservoir acts as a mixing bowl, averaging out normal seasonal fluctuations. The clarity of downstream water also improves as the river's sediment load is trapped by the reservoir. This process may have an adverse effect on the channel below the reservoir both because the streambed erodes and because the natural environment that contributed to the downstream wildlife and vegetation is disturbed. It no longer derives benefits from the annual flooding cycle of sediment-laden water. Similarly, at the upstream end of the reservoir, the capacity of the inlet channel is decreased by sedimentation.

LAKE DIEFENBAKER is a good example of a multi-use reservoir. After 100 years of proposals and plans the project was approved in 1958. On 25 May 1959 an inaugural ceremony marked the start of construction on the project's main dam, GARDINER DAM on the South Saskatchewan River. This dam, and the Coteau Creek generating station, designed to develop hydroelectric power at the damsite, were completed in 1967. A second dam was built on the QU'APPELLE RIVER to keep water from escaping down that stream. In the spring of 1968 Lake Diefenbaker filled to operating capacity and the first power was generated. In 1970, the reservoir filled to its full capacity for the first time.

At its full supply level the dam controls a reservoir that stores a total of 9.4 billion m3, but only 4 billion m3 of this is live capacity. The remaining 5.4 billion m3 is dead storage, and cannot be used for power or to release water downstream. The maximum depth of water at the dam is 58 m and the reservoir created by Gardiner Dam is 225 km long with a shoreline of 760 km. The spillway is 161 m wide and can pass, at its peak capacity, 7500 m3 of water per second. In an average year 250 000 m3 of water is lost to evaporation from the lake's surface area of 430 km2.

Lake Diefenbaker water is also used for irrigation purposes. In 1986 some 16 000 ha were irrigated, and by 1997 that figure had increased to 43 000 ha. It is estimated that, by the year 2030, 115 000 ha will be irrigated. The Coteau Creek generating station develops an average of 775 million kWh each year, and the downstream hydropower stations in the basin benefit directly from the flow regulation of Lake Diefenbaker. For instance, the Squaw Rapids Plant on the Saskatchewan River near Nipawin, Sask, is now able to generate 40% of its power in the winter (double the previous amount) owing directly to improved outflow from Lake Diefenbaker. Water from the lake is also used to supply 10 potash mines, 11 communities and a series of recreational and wildlife projects using water diverted from the lake.

The lake itself has become a water-based resort area, creating a major recreational lake in a traditionally dry area of Saskatchewan, and waters from this lake are annually diverted into the Qu'Appelle River system to augment the municipal water supplies of Moose Jaw and Regina and to improve the quality and quantity of water in the Qu'Appelle River basin.

12

DAMS

An arch-gravity dam, curved-gravity dam or arched dam is a dam with the characteristics of both an arch dam and a gravity dam. It is a dam that curves upstream in a narrowing curve that directs most of the water against the canyon rock walls, providing the force to compress the dam. It combines the strengths of two common dam forms and is considered a compromise between the two. A typical example is the Hoover Dam. A gravity dam requires a large volume of internal fill. An arch-gravity dam can be thinner than the pure gravity dam and requires less internal fill.

Arch-gravity dams are massive dams of reinforced concrete that resist the thrust of water by their weight pushing down using the force of gravity. A gravity dam is constructed so that the dam's massive weight resists the pressure of the water against it.

At the same time an arch-gravity dam incorporates the arch's curved design that is so effective in deflecting the water in narrow, rocky locations where the gorge's side are of hard rock and the water is forced into a narrow channel. Therefore the span needed for the dam is narrow; the dam's curved design effectively holds back the water in the reservoir using a lesser amount of construction material.

When properly situated on an appropriate site, the arch-gravity dam, combining the load resisting qualities of both a gravity dam and an arch dam, inspires the most confidence in the public at large because of its appearance of massive permanence. However, curving a gravity dam may make it look stronger, but some of this effect may be psychological factors in operation in making a choice of dam style.

GRAVITY DAM

The gravity dam is a massive structure made of masonry or concrete with earth and rock fill. Gravity dams use their own weight to resist opposing forces and as such require a hard bedrock foundation. They rely on their great weight and size for stability. The gravity dam is the most commonly built dam in the world. One of the reasons for this is the simplicity of design. Unfortunately it requires great quantities of material to construct.

Arch dam

An arch dam is a thin, curved concrete or masonry dam structure, with the concave side of the curve downstream. Therefore, the force of the water against the canyon wall squeezes the arch, compressing and strengthening its structure and pushing it into the ground. An arch dam is a good dam type for a narrow gorge in a mountainous area with steep walls of rock. It generally has steel rods or prestressed steel cable reinforcements and therefore requires less concrete than a gravity or arch-gravity dam. However, the bedrock in the foundation and abutments must be sound to withstand the pressure of the water.

Arch dams with more than one contiguous arch or plane are described as multiple arch dams. A double arch dam has two contiguous arches. A dam that is curved in both its horizontal and vertical planes may be called a dome dam.

Masonry dams are dams made out of masonry; mainly stone and brick. They are either the gravity or the arch type. The largest masonry dam of the world is Nagarjunasagar Dam in India.

Nagarjuna Sagar Dam

Nagarjuna Sagar Dam is the world's largest masonry dam built across Krishna River in Nagarjuna Sagar, Nalgonda District of Andhra Pradesh, India. It is downstream to the Nagarjuna Sagar reservoir with a capacity of up to 11,472 million cubic metres which is the world's largest man-made lake with a concrete wall of that measures 6 ft (1.8 m). thick. The dam is 490 ft (150 m). tall and 16 km long with 26 gates which are 42 ft (13 m). wide and 45 ft (14 m). tall.

It is one of the earliest irrigation and hydro-electric projects in India. The dam provides irrigation water to the Nalgonda District, Prakasam District, Khammam District and Guntur District.

The proposal to construct a dam to use the excess waters of the Krishna river was put forward by the British rulers in 1903. Siddeswaram, Hyderabad and Pulichintala were identified as the suitable locations for the reservoirs. The perseverance of the Raja of Muktyala paved way for the site identification, design and construction of the dam.

The dam water was released by the then Prime Minister's daughter, Indira Gandhi in 1967. The construction of the dam submerged an ancient Buddhist settlement, Nagarjunakonda, which was the capital of the Ikshvaku dynasty in the 1st and 2nd centuries, the successors of the Satavahanas in the Eastern Deccan.

Excavations here had yielded 30 Buddhist monasteries, as well as art works and inscriptions of great historical importance. In advance of the reservoir's flooding, monuments were dug up and relocated. Some were moved to Nagarjuna's Hill, now an island in the middle of the reservoir. Others were moved to the mainland.

Effect of the project

The project benefited farmers in the districts of Guntur, Prakasam, Krishna, Nalgonda and Khammam. The right canal (a.k.a Jawahar canal) is 203 km long and irrigates 1.113 million acres (4,500 km^2) of land. The left canal (a.k.a Lalbahadur Shastri canal) is 295 km long and irrigates 0.32 million acres (800 km^2) of land in Nalgonda and Khammam districts of Telangana region. The project transformed the economy of above districts. 52 villages were submersed in water and 24000 people were affected. The relocation of the people was completed by 2007.

Power Generation

The hydroelectric plant has a power generation capacity of 815.6 MW with 8 units (1x110 MW+7x100.8 MW). First unit was commissioned on 7 March 1978 and 8th unit on 24 December 1985. The right canal plant has a power generation capacity of 90 MW with 3 units of 30 MW each. The left canal plant has a power generation capacity of 60 MW with 2 units of 30 MW each.

The dam is constructed on the border of Guntur and Nalgonda districts. The dam also provides drinking water to the Nalgonda town.

Statistics

- Catchment Area : 215000 km^2 (83012 sq mi)
- Masonry dam
  - Spillway of dam : 471 m
  - Non-over flow dam : 979 m

- o Length of Masonry dam : 1450 m
- o Maximum height : 125 m
- ⦿ Earth dam
- o Total Length of Earth dam : 3414 m
- o Maximum height : 28 m

Arch dam

An arch dam is a type of dam that is curved and commonly built with concrete. The arch dam is a structure that is designed to curve upstream so that the force of the water against it, known as hydrostatic pressure, presses against the arch, compressing and strengthening the structure as it pushes into its foundation or abutments. An arch dam is most suitable for narrow gorges or canyons with steep walls of stable rock to support the structure and stresses.

Since they are thinner than any other dam type, they require much less construction material, making them economical and practical in remote areas. The development of arch dams throughout history began with the Romans in the 1st century BC and after several designs and techniques were developed, relative uniformity was achieved in the 20th century.

The first known arch dam, the Glanum Dam, also known as the Vallon de Baume Dam, was built by the Romans in France and it dates back to the 1st century BC. The dam was about 12m high and 18m in length. Its radius was about 14m and it consisted of two masonry walls. The Romans built it to supply nearby Glanum with water.

The Monte Novo Dam in Portugal was another early arch dam built by the Romans in 300 AD. It was 5.7m high and 52m long with a radius of 19m. The curved ends of the dam met with two winged walls that were later supported by two buttresses. The dam also contained two water outlets to drive mills downstream.

The Dara Dam was another arch dam built by the Romans in which the historian Procopius would write of its design: "This barrier was not built in a straight line, but was bent into the shape of a crescent, so that the curve, by lying against the current of the river, might be able to offer still more resistance to the force of the stream."

The Mongolians also built arch dams in modern-day Iran. Their earliest was the Kebar Dam built around 1300 which was 26m high, 55m long and had a radius of 35m. Their second dam was built around 1350 and is called the Kurit

Dam. After 4m was added to the dam in 1850, it became 64m tall and remained the tallest dam in the world until the early 20th century. The Kurit Dam was of masonry design and built in a very narrow canyon. The canyon was so narrow that its crest length is only 44% of its height. The dam is still erect even though part of its lower downstream face fell off.

The Elche Dam in Elche, Spain was a post-medieval arch dam built in the 1630s by Joanes del Temple and the first in Europe since the Romans. The dam was 26m high, 75m long and had a radius of 62m. This arch dam also rest on winged walls that served as abutments.

In the 20th century, the world's first variable radius arch dam was built on the Salmon Creek near Juneau, Alaska. The Salmon Creek Dam's upstream face bulged upstream which relieved pressure on the stronger curved lower arches near the abutments.

The dam also had a larger toe which off-set pressure on the upstream heel of the dam which now curved more downstream. The technology and economical benefits of the Salmon Creek Dam allowed for larger and taller dam designs. The dam was therefore revolutionary and similar designs were soon adopted around the world particularly by the U.S. Bureau of Reclamation.

Arch dam designs would continue to test new limits and designs such as the double and multiple-curve. The Swiss engineer Alfred Stucky and the U.S. Bureau of Reclamation would develop a method of weight and stress distribution in the 1960s and arch dam construction in the United States would see its last surge then with dams like the 143m double-curved Morrow Point Dam in Colorado. By the late 20th century, arch dam design reached a relative uniformity in design around the world.

Currently, the tallest and largest arch dam in the world is the 272 m Inguri Dam in Georgia which was completed in 1980. The longest multiple arch with buttress dam in the world is the Daniel-Johnson Dam in Québec, Canada. It is 214 m high and 1,314 meters long across its crest. It was completed in 1968 and put in service in 1970. China is currently constructing the Jinping 1 Dam in Sichuan which when complete will be world's tallest arch dam at 305m.

Design

Arch dams require a high level of stress and force analysis in order to create a sufficient design. The main force against an arch dam is the hydrostatic pressure provided by the reservoir behind it, uplift which is water pressure beneath the dam, the weight of the dam itself and finally all the forces combined. Other forces that affect a dam include, but are not limited to, temperature, chemical reactions, settling, silt accumulation and earthquakes.

Most often, the arch dam is made of concrete and placed in a "V" shaped valley. The foundation or abutments for an arch dam must be very stable and proportionate to the concrete. There are two basic designs for an arch dam: constant radius which have constant radius of curvature and variable radius dams which have both upstream and downstream curves that systematically decrease in radius below the crest.

A dam that is double-curved in both its horizontal and vertical planes may be called a dome dam. Arch dams with more than one contiguous arch or plane are described as multiple-arch dams. Early examples include the Roman Esparragalejo Dam and Muro Dam with later examples such as the Daniel-Johnson Dam (1968) and Itaipu Dam (1982). At an increased cost, an arch dam can have steel rods or pre-stressed steel cable reinforcements and therefore require less concrete than does a gravity dam or arch-gravity dam. As arch dams are commonly constructed in remote mountainous areas, less material is beneficial when it has to be delivered with difficulty to the site. However, as a result of the failure of the Gleno Dam shortly after it was constructed in 1923, the construction of new multiple arch dams has become less popular.

Arch dams are generally classified as thin, medium or thick. If width of an arch dam's base is 2/10 of the dam's height or less it is thin. Any base and height ratio greater than 2/10 but less than 3/10 is considered medium. Thick is a ratio of 3/10 or greater. Additionally, an arch dam's base is historically just as thick as the crest but often it is twice as thick as the crest. Contraction joints are normally placed every 20 m in the arch dam and are later filled with grout after the control cools and cures.

Examples of arch dams

- Inguri Dam
- Buchanan Dam (example of Multiple-Arch type)
- Glen Canyon Dam
- Hoover Dam
- Idukki Dam
- Karun-3 dam
- Daniel-Johnson Dam

Daniel-Johnson Dam

The Daniel-Johnson Dam, formerly known as Manic-5, is a multiple arch buttress dam on the Manicouagan River which creates Manicouagan Reservoir. The dam is composed of 14 buttresses and 13 arches and is 214 km (133 mi) north of Baie-Comeau in Quebec, Canada. The dam was constructed between 1959 and 1970 for the purpose of hydroelectric power production and supplies water to the Manic-5 and Manic-5-PA power houses. The dam is 214 m (702 ft) tall, 1,314 m (4,311 ft) long and contains 2,200,000 m3 (2,880,000 cu yd) of concrete, making it the largest dam of its type in the world.

The dam was named after Daniel Johnson, Sr., the 20th Premier of Quebec who was responsible for starting the project. Johnson died on 26 September 1968, on the day he was to preside over the scheduled inauguration of the dam. The facility is owned and operated by Hydro-Québec.

During the summers of 1919 and 1920, hydrological studies were conducted on the Manicouagan and Outardes rivers, flowing into the St. Lawrence near the town of Baie-Comeau. The combined flow was then estimated at 40 million cubic meters, making it one of the largest hydrological systems in Canada. Although harnessing of this potential was considered interesting, its distance from major load centers and the lack of roads in the area were identified as major drawbacks. In addition, building dams in the wilderness was considered too costly.

After the Second World War, the discovery of large iron deposits on the North Shore and the increased forestry activity led to a rapid development of the region. The area's largest cities, Sept-Iles and Baie-Comeau, were now linked to the rest of the province by a road. At the same time, industrial development in southern Quebec required a larger electric supply. Improvements in long distance electric transmission technologies were also a consideration at the time. Significant advances in the field, including the construction of two 315-kilovolt lines, between the Bersimis complex, west of the Manicouagan system, and Montreal, (completed in 1956), lifted another obstacle.

In 1955, Hydro-Québec launched an 5-year extensive assessment of the Manicouagan's suitability. These studies demonstrated the exceptional potential of the river and stressed the advantages of building a multi-dam system in order take full advantage of the terrain and water flows. The data collected at the time was so promising that Hydro-Québec did not wait for the working group to submit their final recommendations.

By 1959, a decision was made and construction of a 210 kilometres (130 mi) access road from Baie-Comeau was started. The original project, the

Manicouagan-Outardes project, included the construction of five dams on the Manicouagan River (Manic-1, Manic-2, Manic-3, Manic-4 and Manic-5) and three on the Outardes River (Outardes-2, Outardes-3 and Outardes-4). However, a miscalculation prevented the construction of the Manic-4 dam and powerhouse because engineers realized early on that it would encroach on the Manic-3 reservoir.

In 1959, the Premier of Quebec, Maurice Duplessis, wanted a U.S. firm to construct the dam but Quebec's Minister of Water Resources at the time, Daniel Johnson was against the idea. Johnson believed that the skill of a Canadian firm and workers was well suited for the construction of such a complex dam. Duplessis' successor in 1959, Paul Sauvé agreed with Johnson and he continued to pave the way for the dam. Eventually, engineers would choose André Coyne's design for a multiple arch buttress dam as the most suitable and economical.

CONSTRUCTION

Construction on the dam's support facilities began in 1959 and the river diversion tunnels and foundation preparation commenced in 1960. To divert the Manicoagan River, workers blasted and dug two 2,000 ft (610 m) long and 45 ft (14 m) diameter tunnels through the gorge's solid granite west wall.

Workers used "jumbo" drilling platforms and the tunnels progressed 14 ft (4.3 m) each shift. To facilitate the river's diversion, two cofferdams were built. The first was a concrete arch upstream from the dam site which would block the river, forcing it into the diversion tunnels. The second cofferdam was downstream and prevented water from flowing back into the construction site. Each cofferdam was built on alluvial deposits (loose soil) so they needed to be watertight which was accomplished with either a grout curtain or deep supporting piles.

Once the river was diverted, workers pumped the remaining water out between the two cofferdams in order to prepare the site for construction. Once drained, workers excavated the alluvial deposits between the cofferdams creating a long pit that was 150 ft (46 m) deep at the center. This pit was then filled with concrete in the summer of 1962. To prevent seepage in the dam's foundation, a grout curtain was injected in the bedrock and a drainage network with 2,400 ft (730 m) of tunnels was constructed just downstream of the dam to collect water that may seep through.

After the extensive preparatory works were complete, the first concrete for the dam was poured on October 3, 1962. Concrete was poured day and night but was halted during the winter because of freezing temperatures. To organize the pouring, the dam was split into 45 ft (14 m) plots and each was raised about 5 or 6 ft. at a time. Workers had about 150 days - before seasonal flooding - to construct the dam to a height of at least 250 ft (76 m). Before flooding began, engineers planned to seal the diversion tunnels and begin filling the reservoir. To speed the pouring process, concrete was poured in casts by buckets that moved along three cableways suspended above the construction site. The deadline was met within the 150 days and a total of 1,000,000 cu yd (760,000 m3) of concrete was poured. The dam was eventually completed in 1968.

On September 25, 1968, the government-owned utility organized a ceremony to mark the completion of the Manicouagan-5 dam. Hundreds of dignitaries, politicians, utility executives, financiers, engineers and journalists were ferried by plane from Montreal, Quebec City and New York to the work-site to attend a banquet and a plaque unveiling ceremony.

Among the guests were Quebec Union Nationale Premier, Daniel Johnson Sr., his predecessor, Jean Lesage, and René Lévesque, the former Hydraulic Resources minister responsible for the consolidation of all investor-owned utilities into Hydro-Québec. Photographs taken at the banquet show the three men were in excellent spirits, holding hands and smiling, although relations between the Liberal leader and his former cabinet minister were strained by Lévesque's recent defection to the Mouvement Souveraineté-Association, a precursor of the Parti Québécois.

In his authorized biography, Hydro-Québec executive, Robert A. Boyd, recalls being woken up at 6 a.m. the next morning by his boss, Roland Giroux. "I've got bad news, Robert...", said Giroux, adding that he just found the Premier lying dead in his bed. Johnson's sudden demise sent shock waves at the work-site and across the province and the dedication ceremony was quickly canceled.

On September 26, 1969, a year to the day after Johnson's death, the new Premier Jean-Jacques Bertrand accompanied by Johnson's widow and children, unveiled two plaques and officially dedicated the dam after his predecessor. Both plaques now sit side by side at the top of the complex.

FACILITIES

Dam

The design was chosen for strength and economical reasons as it used less concrete or material than a gravity or embankment dam. The Daniel-Johnson Dam is a 214 m (702 ft) tall and 1,312 m (4,304 ft) long multiple-arch buttress dam. Of the dam's 14 total buttresses, the two that form the center arch are 530 ft

(160 m) apart at their base while the others are about 250 ft (76 m) apart. At its thickest point, the center, the dam is 22.5 m (74 ft) wide while the crest can reach about 3 m (10 ft) wide. Pressure from water behind the dam is transferred from the dam's arches to its buttresses and lastly into the ground or its foundation.

The dam was constructed with a high-quality concrete designed to withstand constant thawing and freezing associated with its environment. To further help the structure cope with the climate, engineers placed one inch steel reinforcing bars within the upstream and downstream faces of the dam. The concrete's strength in compression was initially 4,500 lbs. per square inch to meet an estimated 1,500 PSI within the structure. The upstream face of the dam was also coated in asphalt for protection against water. Despite the strength of the dam's concrete, two parallel inclined cracks were discovered on one of the arches shortly after construction.

Reservoir

The dam impounded the Manicougan River, which filled the Earth's fifth largest confirmed impact crater, the Manicouagan crater, creating Manicouagan Reservoir. Coincidentally, the reservoir itself is the fifth largest in the world as well. The reservoir has a maximum depth of 350 m (1,150 ft) feet, mean depth of 85 m (279 ft) and contains 142 km3 (34 cu mi) of water. While draining an area of 29,241 km (18,170 mi), it has a surface area of 1,950 km2 (750 sq mi) and shoreline of 1,322 km (821 mi).

The reservoir is a well known area for Atlantic salmon, lake trout and northern pike fishing, although tall trees flooded during the impoundment have not decomposed due to a lack of oxygen, which can sometimes interfere with the sport.

The priming of the reservoir also created a large artificial island in the center of the Manicouagan reservoir by merging two crescent-shaped lakes: Mouchalagane Lake on the western side and Manicouagan Lake on the eastern side. Covering an area of 2,020 km2 (780 sq mi), René-Levasseur Island is considered to be the second largest island in the world located in a lake, in terms of area (the largest is Manitoulin Island in Lake Huron).

The island was named after René Levasseur, the chief engineer responsible for the construction of the Daniel-Johnson dam. Levasseur died at the age of 35, only days before the dam's inauguration.

Powerhouses

The dam fuels two powerhouses, the Manic-5 and Manic-5-PA. The first powerhouse consists of eight Francis turbines, capable of producing up to

1,528 MW of power, which went online in 1970. The second powerhouse, the Manic-5-PA (PA stands for puissance additionnelle or additional power), was commissioned in 1989, and consists of four Francis turbines of 1,064 MW in total installed capacity.

The designers of the Manic-5 decided on an above ground power house that was downstream of the dam for safety and cost. The intake was built on the east side of the dam and supplies two 3,400 ft (1,000 m), long 36 ft (11 m) diameter concrete-lined penstocks (tunnels). Just before reaching the power house and its eight turbines, each penstock splits into four branches.

The power house is about 2,500 ft (760 m) downstream of the dam and utilizes two surge tanks for sudden rises in water pressure from the two penstocks. Each surge tank has a 80 ft (24 m) diameter expansion chamber and is about 40 ft (12 m) higher than the actual dam structure. The surge tanks protect the penstocks and turbines from water hammer which would occur if the turbine gates were quickly closed and water pressure suddenly increases.

Bligh's Creep Theory

Design of impervious floor for sub surface flow: Iit is directly depended on the possibilities of percolation in porous soil on which the floor (apron) is built. Water from upstream percolates and creeps (or travel) slowly through weir base and the subsoil below it. The head lostby the creeping water is proportional to the distance it travels (creep length)along the base of the weir profile. The creep length must be made as big as possible so as to prevent the piping action. This can be achieved by providing deep vertical cut-offs or sheet piles.

According to Bligh's theory, the total creep length for first drawing: $L = B$ and for second drawing: $L = B + 2(d1 + d2 + d3)$

If H is the total loss of head, then the loss of head per unit length of the creep shall be

Bligh called the loss of head per unit length of creep as Percolation coefficient. The reciprocal, (L/H) of the percolation coefficient is known as the coefficient of creep C.

Assumptions

- Hydraulic slope or gradient is constant throughout the impervious length of the apron.
- The percolating water creep along the contact of the base profile of the apron with the sub soil losing head enroute, proportional to length of its

travel. The length is called creep length. It is the sum of horizontal and vertical creep.

- Stoppage of percolation by cut off (pile) possible only if it extends up to impermeable soil strata.

Design criteria:

Safety against piping:

The length of creep should be sufficient to provide safe hydraulic gradient according to the type of soil.

Safe creep length = L = CH, C = 1/c

Safety against uplift pressure:

Let h' = uplift pressure head at any point of apron (Hydraulic gradient line above the bottom of floor)

The uplift pressure = wh' where w = density of water. If t = thickness of floor at the point, l = specific gravity for floor material. Then, downward force per area (resisting force) = t.w.e or wh' = t.w.e

For portion of floor upstream of barrier only nominal thickness need to be provided since the weight of water will counterbalance the uplift pressure.

A certain minimum length of impervious floor is always necessary to the downstream of the barrier (thickness of downstream floor for worst condition)

Limitations of Bligh's theory

1. This theory made no distinction between horizontal and vertical creep.
2. Did not explain the idea of exit gradient - safety against undermining cannot simply be obtained by considering a flat average gradient but by keeping this gradient will be low critical.
3. No distinction between outer and inner faces of sheet piles or the intermediate sheet piles, whereas from investigation it is clear, that the outer faces of the end sheet piles are much more effective than inner ones.
4. Losses of head does not take place in the same proportions as the creep length. Also the uplift pressure distribution is not linear but follow a sine curve.
5. In case of two piles the width between should be greater than twice the head or the piles are not effective.

Khosla's theory

After studying a lot of dam failure constructed based on Bligh's theory, Khosla came out with the following;

Following are some of the main points from Khosla's Theory

- From observation of Siphons designed on Bligh's theory, by actual measurement of pressure, with the help of pipes inserted in the floor of two of the siphons?
- Does not show any relationship with pressure calculated on Bligh's theory. This led to the following provisional conclusions:
- Outer faces of end sheet piles were much more effective than the inner ones and the horizontal length of the floor.
- Intermediated piles of smaller length were ineffective except for local redistribution of pressure.
- Undermining of floor started from tail end.

For case of two dimensional flows under a straight floor where:

Thus for the first flow line AB which touches the outline of the floor, the pressure can be obtained by putting different values of x in equation. Fig shows the pressure distribution diameter both by equation 4 as well as Bligh's Theory.

From the fig the following conclusions can be drawn

Slope of Pressure diagram: At A and B in infinite, hence the floor at A will be theoretically infinite acting downward and that at B will also be infinite acting upward. This will cause sand boiling and hence the floor should be depressed or cut off should be provided at the downstream end.

Composite profile

The following specific causes of general form were considered.

- Straight horizontal flow of negligible thickness with pile at either end, upstream or at downstream end.
- Straight horizontal floor of negligible thickness with pile at some intermediate point.
- Straight horizontal floor, depressed below the bed, but with no cut off.

Method of independent variable:

- Most designs do not confirm to elementary profiles (specific cases). In actual cases we may have a number of piles at upstream level, downstream level and intermediate points and the floor also has some thickness.
- Khosla solved the actual problem by an empirical method known as method of independent variables.
- This method consists of breaking up a complex profile into a number of simple profiles, each of which is independently amiable to mathematical treatment. Then apply corrections due to thickness of slope of floor.
- As an example the complex profile shown in fig is broken up to the following simple profile and the pressure at Key Points obtained.
- Straight floor of negligible thickness with pile at upstream ends.
- Straight floor of negligible thickness with pile at downstream end.
- Straight floor of negligible thickness with pile at intermediate points.
- The pressure is obtained at the key points by considering the simple profile.

For the determination of seepage below the foundation of hydraulic structure developed the method of independent variable.

In this method, the actual profile of a weir which is complex, is divided into a number simple profiles, each of which cab be solved mathematically without much difficulty. The most useful profile considered are:

A straight horizontal floor of negligible thickness provided with a sheet pile at the upstream end or a sheet pile at the downstream end.

ii) A straight horizontal floor depressed below the bed, but without any vertical cut-off.

1. A straight horizontal floor of negligible thickness with a sheet pile at some

Intermediate point

The mathematical solution of the flow-nets of the above profiles have been given in the form of curves. From the curves, percentage pressures at various key points E, C, E1, C1 etc) be determined. The important points to note are:

1. Junctions of pile with the floor on either side{E, C (bottom), E1, C1 (top)}
2. Bottom point of the pile (D), and

3. Junction of the bottom corners (D, D') in case of depressed floor

The percentage pressures at the key points of a simple forms will become valid for any complex profile, provided the following corrections are effected:

- correction for mutual interference of piles
- correction for the thickness of floor
- correction for slope of the floor.

Correction for Mutual Interference of Piles

Let b1 = distance between the two piles 1 and 2, and

D = the depth of the pile line (2), the influence of which on the neighbouring pile (1) of depth d must be determined

b = total length of the impervious floor

c = correction due to interference.

The correction is applied as a percentage of the head

This correction is positive when the point is considered to be at the rear of the interfering pile and negative for points considered in the forward or flow direction with the interfering pile.

Correction for Floor Thickness

Standard profiles assuming the floors as having negligible thickness. Hence the values of the percentage pressures computed from the curves corresponds to the top levels (E1*, C1*) of the floor. However, the junction points of the floor and pile are at the bottom of the floor (E1, C1)

The pressures at the actual points E1 and C1 are interpolated by assuming a straight line variation in pressures from the points E1* to D1 and from D1 to C1

The corrected pressures at E1 should be less than the computed pressure t E1*. Therefore the correction for the pressure at E1 will be negative. And so also is for pressure at C1.

BIBLIOGRAPHY

- Froriep, Siegfried (1986): "Ein Wasserweg in Bithynien. Bemühungen der Römer, Byzantiner und Osmanen", Antike Welt, 2nd Special Edition, pp. 39-50
- Grewe, Klaus (2008): "Tunnels and Canals", in: Oleson, John Peter (ed.): The Oxford Handbook of Engineering and Technology in the Classical World, Oxford University Press, pp. 319-336, ISBN 978-0-19-518731-1
- Moore, Frank Gardner (1950): "Three Canal Projects, Roman and Byzantine", American Journal of Archaeology, Vol. 54, No. 2, pp. 97-111
- Schörner, Hadwiga (2000): "Künstliche Schiffahrtskanäle in der Antike. Der sogenannte antike Suez-Kanal", Skyllis, Vol. 3, No. 1, pp. 28-43
- Serban, Marko (2009): "Trajan's Bridge over the Danube", The International Journal of Nautical Archaeology, Vol. 38, No. 2, pp. 331-342
- Tudor, D. (1974): Les ponts romains du Bas-Danube, Bibliotheca Historica Romaniae Études, Vol. 51, Bucharest: Editura Academiei Republicii Socialiste România, pp. 47-134
- White, K. D. (1984): Greek and Roman Technology, London: Thames and Hudson, pp. 110-112; 227-229, table 6
- Wikander, Charlotte (2000): "Canals", in Wikander, Örjan (ed.): Handbook of Ancient Water Technology, Technology and Change in History, Vol. 2, Leiden: Brill, pp. 321-330, ISBN 90-04-11123-9

- Burton, Anthony (1995) [1989], The Great Days of the Canals, Twickenham: Tiger Books, ISBN 1-85501-695-8
- Calvert, Roger (1963), Inland Waterways of Europe, George Allen and Unwin
- Edwards-May, David (2008), European Waterways - map and concise directory, 3rd edition, Euromapping
- Hadfield, Charles (1986), World Canals: Inland Navigation Past and Present, David and Charles, ISBN 0-7153-8555-0
- Needham, J (1971), Science and Civilisation in China, C.U.P. Cambridge
- Rodda, J. C. (2004), The Basis of Civilization - Water Science?, International Association of Hydrological Sciences
- Blair, John (ed.) (2007). Waterways and Canal-building in Medieval England. Oxford: Oxford University Press. ISBN 978-0-19-921715-1.
- Broadbridge, S.R. (1974). The Birmingham Canal Navigations. Volume 1: 1768-1846. Newton Abbot: David & Charles. ISBN 0-7153-6381-6.
- Burton, Anthony (1995). The Great Days of the Canals. London: Tiger Books International. ISBN 1-85501-695-8.
- Burton, Anthony (1983). The Waterways of Britain: A Guide to the Canals and Rivers of England, Scotland and Wales. London: Willow Books, William Collins and Sons & Co Ltd. ISBN 0-00-218047-2.
- Hadfield, Charles (1966). The Canals of the West Midlands. Newton Abbot: David & Charles. ISBN 0-7153-4660-1.
- Hadfield, Charles (1981). The Canal Age (Second ed.). David & Charles. ISBN 0-7153-8079-6.
- Lindsay, Jean (1968). The Canals of Scotland. Newton Abbot: David & Charles. ISBN 0-7153-4240-1.
- Malet, Hugh (1961/1990). Bridgewater: The Canal Duke 1736-1803, 3rd rev ed, paperback. Nelson, UK: Henton Publishing Co. ISBN 0-86067-136-4.
- Paget-Tomlinson, E. (2006) The Illustrated History of Canal & River Navigations: Landmark Publishing Ltd ISBN 1-84306-207-0
- Palmer (chairman), Robert (1955). "Canals and Inland Waterways, Report of the Board of Survey". British Transport Commission.

- "Reader's Digest Library of Modern Knowledge". London: Reader's Digest. 1978..
- Rolt, L.T.C. (1944). Narrow Boat. London: Eyre Methuen. ISBN 0-413-22000-1.
- Rolt, L.T.C. (1950). The Inland waterways of England. London: George Allen and Unwin Ltd. ISBN 0-04-386003-6.
- Roger Squires (2008). Britain's restored canals. Landmark Publishing. ISBN 978-1-84306-331-5.
- Thompson, Hubert Gordon (1904). The Canal System of England. London: T. Fisher Unwin.
- Bliault, Charles; Jonas, Martin; The North of England P&I Association (2016). Bulk Cargoes: A Guide to Good Practice (First ed.). UK: The North of England P&I Association. p. 280. ISBN 978-0-9574936-3-6. ASIN 0957493630.
- George, William (2005). Stability and Trim for the Ship's Officer. Centreville, MD: Cornell Maritime Press. ISBN 978-0-87033-564-8.
- Hayler, William B.; Keever, John M. (2003). American Merchant Seaman's Manual. Cornell Maritime Pr. ISBN 0-87033-549-9.
- Albert Allen Bartlett, Arithmetic, Population, and Energy: The Forgotten Fundamentals of the Energy Crisis, (1978)
- Alan Weisman. Countdown: Our Last, Best Hope for a Future on Earth? Little, Brown and Company, (2013) ISBN 0316097756
- Andrew Mason, Population change and economic development in East Asia: Challenges met, opportunities seized(2001)
- Barry Commoner, Making Peace with the Planet (1990)
- Ben J. Wattenberg, The Birth Dearth (1989) Fewer: How the New Demography of Depopulation Will Shape Our Future, (2005)
- CK-12 Agriculture and human population growth
- Donella Meadows, Jorgen Randers, Dennis Meadows, Limits to Growth: The 30-Year Update (Paperback) (2004)
- Edwards, Paul Geoffrey; Miller, Clark A. (2001). Changing the atmosphere: expert knowledge and environmental governance. Cambridge, Mass: MIT Press. ISBN 0-262-63219-5.

- Emanuel K (August 2005). "Increasing destructiveness of tropical cyclones over the past 30 years" (PDF). Nature 436(7051): 686–8. Bibcode:2005Natur.436..686E. PMID 16056221.
- F. L. Lucas, The Greatest Problem (1960); an early wake-up call on over-population, by a distinguished Cambridge academic
- Garrett Hardin, Living Within Limits, (1995) reprint
- Herman Daly, Ecological Economics and the Ecology of Economics (1999)
- IPCC AR4 SYR (2007). "Summary for Policymakers". In Core Writing Team; Pachauri, R.K; and Reisinger, A. Climate Change 2007: Synthesis Report. Contribution of Working Groups I, II and III to the Fourth Assessment Report of the Intergovernmental Panel on Climate Change. IPCC. ISBN 92-9169-122-4..
- Joel E. Cohen, How Many People Can the Earth Support? (1996)
- Julian Lincoln Simon, The Ultimate Resource 2, (1998)"
- McKibben, Bill (2011). The Global Warming Reader. New York, N.Y.: OR Books. ISBN 978-1-935928-36-2.
- Moseley, W.G. 2007. "A population remedy is right here at home: U.S. overconsumption is a bigger issue than fertility." The Philadelphia Inquirer. 11 July.
- Paul R. Ehrlich, The Population Bomb, (1968) The Population Explosion, (1990) The Population Bomb, (1995) reprint
- Ruddiman, W. F. (2003). "The anthropogenic greenhouse era began thousands of years ago". Climate Change 61 (3): 261–293. doi:10.1023/B:CLIM.0000004577.17928.fa.
- Ruddiman, W. F., Vavrus, S. J. and Kutzbach, J. E. (2005). "A test of the overdue-glaciation hypothesis". Quaternary Science Reviews 24 (11): 1. Bibcode:2005QSRv...24....1R.
- Safa Motesharrei, et al. A Minimal Model for Human and Nature Interaction. University of Maryland, (2012).

INDEX

9 789360 840082

Printed at Repro India Ltd.